MOLECULAR DIVERSITY IN DRUG DESIGN

MOLECULAR DIVERSITY IN DRUG DESIGN

Edited by

PHILIP M. DEAN

and

RICHARD A. LEWIS

SPRINGER-SCIENCE+BUSINESS MEDIA, B.V.

A C.I.P. Catalogue record for this book is available from the Library of Congress.

ISBN 978-94-017-3831-6   ISBN 978-0-306-46873-5 (eBook)
DOI 10.1007/978-0-306-46873-5

Printed on acid-free paper

Contents

Contributors

Michael Almstetter
Morphochem AG,
Am Klopferspitz 19,
82152 Martinsried, Germany
Per M. Andersson
Research Group for Chemometrics
Department of Organic Chemistry,
Institute of Chemistry,
Umeå University,
SE-904 87 Umeå, Sweden
Keith Davies
Department of Chemistry
University of Oxford, UK
EKDavies@compuserve.com
Philip M. Dean
Drug Design Group
Department of Pharmacology
University of Cambridge, UK
pmd1@ddg.phar.cam.ac.uk
Valerie J. Gillet
University of Sheffield
Western Bank
Sheffield S10 2TN, United Kingdom
V.Gillet@sheffield.ac.uk

Richard A. Lewis
Eli Lilly & Co. Ltd
Lilly Research Centre
Windlesham
Surrey GU20 6PH UK
lewis_richard@lilly.com
Anna Linusson
Research Group for Chemometrics
Department of Organic Chemistry,
Institute of Chemistry,
Umeå University,
SE-904 87 Umeå, Sweden
Torbjörn Lundstedt
Structure Property Optimization Center,
Pharmacia & Upjohn AB,
SE-751 82 Uppsala, Sweden
Jonathan S. Mason
Bristol-Myers Squibb
PO Box 4000
Princeton NJ08543 USA
masonj@bms.com
Christopher G. Newton
Dagenham Research Centre
Rhone-Poulenc Rorer
Rainham Road South
Dagenham, Essex RM10 7XS UK
Christopher.Newton@rp-rorer.co.uk
Bo Nordén
Medicinal Chemistry,
Astra Hässle AB,
SE-43183 Mölndal, Sweden
Diana C. Roe
Sandia National Labs,
Mail Stop 9214, P.O. Box 969,
Livermore, CA 94551 USA
dcroe@ca.sandia.gov
Michael Sjöström
Research Group for Chemometrics,
Department of Organic Chemistry,
Institute of Chemistry,
Umeå University,
SE-904 87 Umeå, Sweden

James Smith
Drug Design Group
Department of Pharmacology,
University of Cambridge, UK
Svante Wold
Research Group for Chemometrics,
Department of Organic Chemistry,
Institute of Chemistry,
Umeå University,
SE-904 87 Umeå, Sweden
svante.wold@chem.umu.se
Lutz Weber
Morphochem AG,
Am Klopferspitz 19,
82152 Martinsried, Germany
lutz.weber@morphochem.de
Peter Willett
Department of Information Studies,
University of Sheffield,
Western Bank,
Sheffield S10 2TN, UK.
p.willett@sheffield.ac.uk
Catherine White
Oxford Molecular Group,
Oxford, UK

Acknowledgements

P.M.D. would like to acknowledge the Wellcome Trust for encouragement and financial support provided over many years; Rhone-Poulenc Rorer have also contributed significant laboratory funding. R.A.L. would like to thank Ann-Marie O'Neill for her patience, and the management at Eli Lilly and Rhone-Poulenc Rorer for providing environments that allowed scientific debate to flourish. We are grateful to Peter Butler, Janet Hoffman and the staff of Kluwer for their help in producing this book; any errors that remain are solely the fault of the editors.

Preface

High-throughput screening and combinatorial chemistry are two of the most potent weapons ever to have been used in the discovery of new drugs. At a stroke, it seems to be possible to synthesise more molecules in a month than have previously been made in the whole of the distinguished history of organic chemistry, Furthermore, all the molecules can be screened in the same short period. However, like any weapons of immense power, these techniques must be used with care, to achieve maximum impact. The costs of implementing and running high-throughput screening and combinatorial chemistry are high, as large dedicated facilities must be built and staffed. In addition, the sheer number of chemical leads generated may overwhelm the lead optimisation teams in a hail of friendly fire. Mother nature has not entirely surrendered, as the number of building blocks that could be used to build libraries would require more atoms than there are in the universe. In addition, the progress made by the Human Genome Project has uncovered many proteins with different functions but related binding sites, creating issues of selectivity. Advances in the new field of pharmacogenomics will produce more of these challenges. There is a real need to make high-throughput screening and combinatorial chemistry into 'smart' weapons, so that their power is not dissipated. That is the challenge for modellers, computational chemists, cheminformaticians and IT experts.

In this book, we have broken down this grand challenge into key tasks. In chapter 1, Smith, Dean and Lewis define in detail many of the key issues in molecular diversity and in the analysis of binding sites, showing how subtle changes in sequence can be translated into features that could drive library design. The next chapter by Newton deals with the considerable logistical and managerial challenges of running combinatorial chemistry and high-throughput screening laboratories, and gives a clear picture of how to obtain the best value from these operations. Chapter 3 by Gillet lays out rigorously the theory underpinning molecular diversity and the design of libraries, followed by a practical demonstration of the theory by Mason in his elegant paper applying 4-centre pharmacophores to the design of privileged libraries. In chapter 5, Weber and Almstetter describe recent advances in methods for dealing with very large libraries, libraries that could never be contemplated without the tools provided by molecular diversity. Again, both the theory and practical examples are given. Next, Willett reviews critically all the current methods for selecting subsets of libraries, providing clear guidance as to the best ways to approach this task. Roe then tackles the particular issue of how to design libraries using the constraints of a protein active site; the prospects offered by the marriage of site-directed drug design and molecular diversity are very exciting. In chapter 8, Davies and White discuss the IT issues created by the large volumes of data that can be created during the design, synthesis and screening of combinatorial libraries. It is very apparent that the ability to store and query large volume of textual, numeric and structural data needs to be seen as the new and required enabling technology if the field is to move forward and realise its promise. Combinatorial libraries can also be small, if the products are expensive or difficult to make. Andersson *et al.* show how chemometrics can be applied to get the most value out of small libraries, using procedures that will be new to most medicinal chemists. Finally, Lewis discusses how to design small and medium-sized libraries to optimise SARs; this is in recognition of the fact that the techniques of combinatorial chemistry are increasingly being used by medicinal chemists during lead optimisation. Each chapter conveys, we hope, how exciting and intellectually challenging the field of molecular diversity in drug design is. We expect the next five years to generate even more advances, to modify the scatterguns of high-throughput screening and combinatorial chemistry into weapons more capable of firing the elusive magic bullets of potent drugs.

Philip Dean

Richard Lewis

July 1999

Chapter 1

Issues in Molecular Diversity and the Role of Ligand Binding Sites

Molecular Diversity and Ligand Binding Sites

JAMES SMITH[1], PHILIP M. DEAN[1] AND RICHARD A. LEWIS[2]

1. Drug Design Group, Department of Pharmacology, University of Cambridge, UK

2. Eli Lilly & Co. Ltd, Lilly Research Centre, Windlesham, Surrey GU20 6PH UK

Key words: receptor binding site, molecular diversity, ligand flexibility, design strategies

Abstract: The role of molecular diversity in the design of combinatorial libraries is discussed with respect to the strategic issues that arise from the sheer numerical scale of combinatorial chemistry and high-throughput screening, and the issues that arise when applying binding site information to the design process. A method for the analysis of binding sites, that can be used to explore the common features and the differences between a set of related binding sites, is presented. The method is applied to the analysis of nucleotide binding sites.

1. ISSUES IN MOLECULAR DIVERSITY

The goal of molecular diversity research is to provide better methods for harnessing the power of combinatorial chemistry and high-throughput screening. Most of the content of the other chapters in this book deals with the diversity within and between sets of small ligand molecules. This paper concentrates on the design of either optimally diverse general libraries of small molecules or focused libraries of small molecules that explore a structure-activity relationship (SAR) provided, for example, by a receptor binding site. The two strands to the design strategies are quite different and have to be considered as distinct research problems. With regard to differences between sites, two further practical problems arise, since most sites within a functional family show similarities due to a common evolutionary pathway. In practice, focused library design could be divided into two categories: design to a general class of sites (for example tyrosine kinases, where only subtle differences between the sites are present); design to a specific site for which complete specificity is required (for example, for a cell-type specific and ligand specific tyrosine kinase). We will first consider briefly some of the general issues of library design and combinatorial chemistry, before turning to a detailed discussion of the challenges of binding site analysis.

1.1 Definitions

What is combinatorial chemistry? One flippant answer might be "A method of increasing the size of the haystack in which you find a needle"[1]. Combinatorial chemistry involves the combination of reagents, according to a synthetic scheme, to generate products with one or more variable R-group positions (figure 1). The upper limit to the size of the combinatorial library that could be generated is given by the product of the number of possible reagents at each of the substituent positions. For example, if a scheme involves three different reagents, and there are 100 possible reagents of each type, then the synthesised combinatorial library would contain 1 million compounds. A library can be built, or enumerated, in the computer, as a precursor to design and/or synthesis; the obvious term for this is a 'virtual library'.

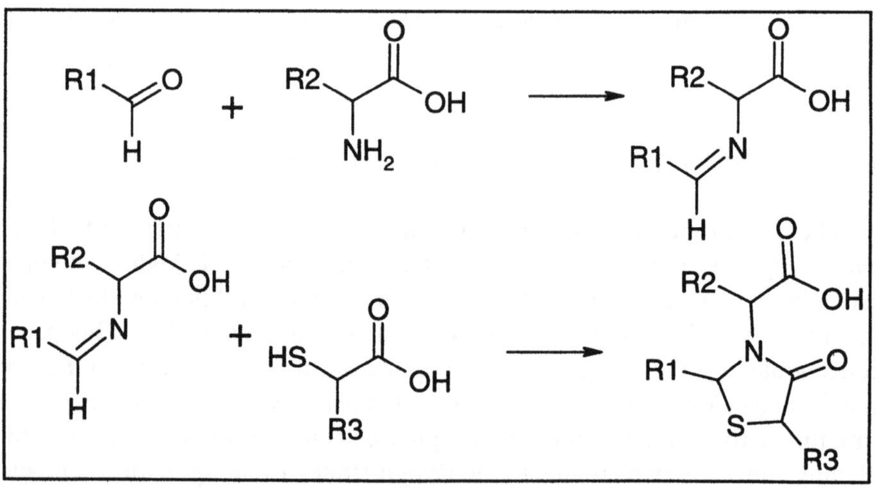

Figure 1. A synthetic scheme that will generate a combinatorial library with three sites of variation.

The definition of what is a small or medium-sized library is fairly arbitrary, depending as it does on working practices within a laboratory. A team which places particular emphasis on purity may find that the rate-limiting step is the purification, rather than the actual synthetic steps. Similar constraints can be imposed by the complexities of the chemistries being attempted, and the final quantities required for screening. For the purposes of this paper, a small library will consist of only a few hundred members, while a medium library may have up to a few thousand members.

1.2 Combinatorial Efficiency

The minimum number of reagents needed to make N products in a k-component reaction is $k \cdot N^{1/k}$. The maximum number is k•N. Design methods that try to use the minimum number of reagents are called 'efficient', whereas those that tend towards larger numbers are termed 'cherry-picking'. The terms are not meant to be derogatory: the key factor in the design should be the exploration or refinement of an SAR, rather than the numbers of reagents used in the synthesis. Against that, it can be quite tedious and time-consuming to make a medium library that has been designed with no heed to efficiency. Thus, medium libraries will tend towards being truly combinatoric, that is, made up of all possible combinations of reagents, while small libraries need not be. This distinction is important, because it changes the design goal of maximising some measure of diversity and combinatorial efficiency to simply maximising diversity. In this latter situation, cherry-picking methods can be used. There is no universal recipe, and each case should be looked at on its own merits.

1.3 Diversity and Similarity

The terms 'similarity' and 'diversity' are very nebulous, as they seem to encompass several different concepts in the literature; Kubinyi has published an interesting polemic on this issue [2]. A narrow definition, which revolves around the context of exploring an SAR, will be employed in this work. Small and medium libraries are made for the purpose of following up leads and exploring their SAR as quickly as possible. The libraries therefore need to be designed around a hypothesis of what makes the leads active, and what factors might contribute to the SAR which have not been explored yet. The library design, and hence the definition of diversity, must therefore vary from case to case. The logical conclusion of this line of argument would be exemplified by the design of peptides to bind to proteases: each design is based around a common core and a common chemistry, but each design will be different, driven by the environment of the different enzymes. One can make some general remarks about which descriptors will probably be important, and this will be covered later. Diversity is therefore the spread of observations in a defined descriptor space, and within defined limits of that space, the descriptors and limits being determined by the nature of the SAR, and the amount of knowledge available about the SAR.

1.4 Work Flows in Combinatorial Chemistry

A typical work flow for the conception, design and synthesis of a library by combinatorial chemistry or rapid parallel synthesis (RPS) is shown in figure 2. The starting point is SAR information and a synthetic plan for making the library. The first phase revolves around testing out the feasibility of the synthetic scheme and gathering information on the reagents available for use. These two processes can impose limits on the virtual library through providing constraints on what reagents will react, what reagents are available in the stockroom or by quick delivery. This leads into the reagent filtering phase, which results in the final set of reagents for enumeration into the virtual library. The next phase is the design phase, which takes input from the SAR and other sources. Closely allied to the design phase is the inspection phase, in which the compounds chosen by the design are eyeballed by experienced medicinal and RPS chemists to make sure that the design meets their expectations. The next stages are synthesis, purification and registration, followed by screening and validation. If a library has been carefully designed according to an explicit hypothesis, then the results from biological screening should serve to test the hypothesis and guide the design of the next library. If the design has not been driven in this way, it will be that much harder to elucidate the SAR information, thus defeating the object of making the library in the first place.

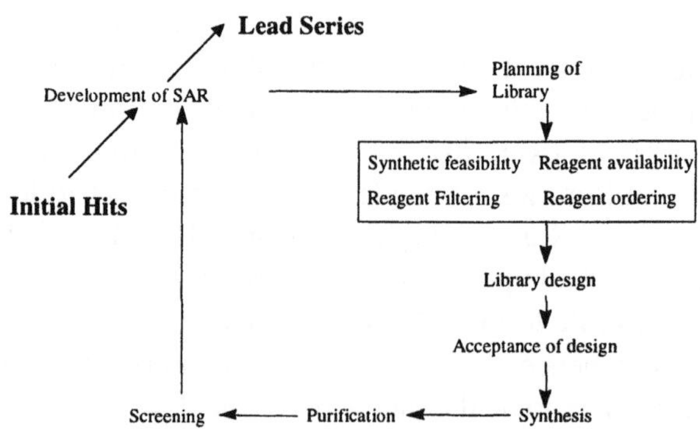

Figure 2. A workflow for the design, preparation and use of a combinatorial library.

1.5 Combinatorial Chemistry and Diversity Analysis - Why bother?

The practise of combinatorial chemistry is a costly business, requiring expensive equipment and skilled chemists, as is the use of drug resources to perform diversity analyses. Any scientists embarking on a project in these fields should ask themselves why they are using these tools in the first place, and how to get the best value from them when they are used. The answer to these questions will depend on whether one is involved in high-throughput screening (HTS), combinatorial chemistry, drug design or management, whether you are part of an academic or industrial group, and how how large or small your organisation is.

The HTS-biased answer is that testing more compounds increases the probability of finding a good lead. 'The more you test, the more you know' [3]. The number of compounds that could theoretically be made even with today's technologies would keep HTS happy for a while. Several authors have asked the question whether the advances made in combinatorial chemistry herald the end of rational drug design. The conclusion of these studies has been that, however fast and efficient the techniques of combinatorial chemistry become, the number of compounds that could be made far outweighs the current capacity to store, screen and validate, so that design has just as much of a role to play as in the situations of chemical scarcity, where each compound is the product of a long individual synthesis. In addition, the logistics of performing these functions at very high throughput rates are beyond all but the largest organisations and infrastructures. A smaller group must work within its limitations, and if those can be expressed in the number of compounds that can be processed, then diversity analysis has a role to play in focusing effort in a productive manner.

The business-oriented philosophy can be expressed as: "It is because the cost of finding leads against a pharmaceutical target, which can then be optimised quickly to a candidate drug has become very expensive, and it has been claimed that considerations of molecular diversity in selecting compounds for screening may reduce this expense." [4]. Managers and project leaders are concerned that the output from combinatorial chemistry should be worth screening, in terms of diversity and in potential ADME (Absorption, Delivery, Metabolism, Excretion) issues.

The chemistry perspective is often driven by what can be made, in an effort to push back the envelope of what can be achieved by the equipment and the chemistry. The product of these endeavours might be a large number of exotic and chemically interesting compounds which have a dim future as drugs, but are good sources of journal papers. In this case, the chemistry is an end in itself, and diversity analysis has no part to play.

The next issue is one of originality, that is, has anyone made these compounds or this library before? The patents issue is perhaps the least well documented, although it is a common issue. The number of reactions that respond well to combinatorial approaches (to yield reasonable amounts of product of > 80% purity) is limited but growing. It is reasonable to assume that if company A uses a literature reaction, then company B will have used the same reaction, possibly to make the same products. The chances of making a unique product are therefore greatly diminished. Unfortunately, there is no fast mechanism for checking for patentability at present, although the work of Barnard and Downs on Markush representations offers a future solution [5].

The drug designer's perspective is towards working out what compounds should be made. In an academic context, this will involve inventing new and general methods for describing molecules, for calculating similarity and diversity, and for testing these methods against some known data set of actives and inactives. This work is extremely valuable, as it moves the science of diversity analysis forward. However, a modeller in the pharmaceutical or agrochemical industries would be more concerned about finding the best particular method for dealing with the project in hand. This is not just a matter of good science, but of timeliness as well. Can the design be done fast enough and well enough to influence chemistry? A useful parallel is structure-based drug design, where calculations of reasonable precision have been of more use historically than more rigourous calculations, which have taken as long and required as many resources as making the compound itself.

Any combinatorial chemistry campaign or design strategy should be geared to finding leads quickly, and furthermore should enable those hit compounds to be turned into a lead series. Ecker and Crooke proposed criteria that define value as regards combinatorial chemistry and diversity analysis [6]:

- Improvements in the quality of drug candidates
  - Rapid Identification of potent leads
  - Rapid movement of leads into the clinic
- Improvements in the quality and performance of drug candidates
  - Major improvements in specificity
  - improvements in ADME
  - low toxicity
- Success where traditional methods have failed.

Two examples spring to mind: peptide libraries can be made quickly, contain large numbers of compounds and have shown value in finding leads quickly. However, it is hard to make a hit from a peptide library into a lead series. Libraries built along a benzodiazepine core also yield a good supply of hits that are easy to optimise into drug-like molecules. Against that, it may be much harder to secure a good patent position, if that matters to one.

The purpose of this chapter (and the ones that follow it) is not to explore the synthetic and technical possibilities of combinatorial chemistry. It is assumed that they are virtually limitless, like chemistry itself. However, we are concerned with the pragmatic application of these methods to whatever ends they are directed, whether it be the production of novel pharmaceutical compounds, or papers in learned journals.

1.6 The Similarity Principle

Medicinal chemistry is a very challenging discipline, based on a large amount of case lore, as often there is neither the resources nor inclination to prove a structure-activity relationship (SAR) conclusively. What has sprung up instead is the similarity principle, which is that similar molecules generally produce similar biological effects. This implies that dissimilar compounds generally produce dissimilar effects. This principle holds only if you have the right method for measuring similarity and dissimilarity. It will break down, if you are not comparing like with like, for example if there are multiple binding modes to the target, or if the biological endpoint is the product of several independent processes which have a different influence on different lead series.

The similarity principle also implies that changes in activity are gradual, if only small changes in molecular structure (implying high similarity to previous compounds) are made. In terms of molecular interactions, we are saying that changes that don't strongly affect the stability of the ligand-receptor complex, the relative population of the binding conformation of the ligand in solution, or the solvation energy are the norm. Medicinal chemistry is littered with examples where a small change in structure leads to large change in activity, both positive and negative. Lajiness has described this phenomenon as an 'activity cliff' in his work on trying to incorporate these observations into SAR models [7]. Despite these problems, the similarity principle is a good starting point for diversity analyses.

The similarity principle allows one to formulate answers to questions as to how to construct a representative sample of a large combinatorial library, or how to design an efficient library to follow up an active lead compound. We can now postulate that a representative sample that contains molecules that are too similar will be wasteful, as the active structure will probably be duplicated. However, for lead follow-up we require the molecules to be quite similar to the lead, but not excessively so. The cliché 'methyl, ethyl, propyl, futile' springs to mind.

This line of thought leads to the question of how much similarity is enough? Patterson *et al.* have proposed a solution in terms of the construction of a 'neighbourhood region' around a molecule defined in descriptor space [8]. This method is akin to sphere-inclusion/exclusion methods, and the diameter of the sphere can be estimated from analysis of other SARs.

1.7 Validation

Validation of diversity space is an unsolved problem, fraught with difficulties. Validation implies test comparison of our theoretical results against some absolute truth, provided by experimental data or by the universe of all possible results. Our stated goal is that design should enhance the process of lead generation and optimisation. It would seem appropriate to use hit rates as a measure of how well our diversity analysis does as compared to chance: "simulated screening". This approach has been investigated by several workers [9]. It assumes that the universe of chemical space can be neatly divided into actives and inactives, according to some biological test. Membership of a set depends upon the threshold defined for activity. Thus, membership of the actives club becomes more exclusive as the threshold is raised and fewer chemical families are able to gain entrance.

A similar idea has been expressed by Cramer and co-workers as trying to find the 'activity island' [10]. It should be noted that this approach does make the implicit assumption that there is reliable information in inactive compounds, an idea that we are not entirely comfortable with.

A key issue in descriptor validation is how to define a reference set which is meant to typify the universal set of actives, and possibly inactives. One approach has been to use the World Drug Index [11] to define the set of active compounds, and the Spresi database [12] to define the inactives. Care has to be taken when using the WDI, as it contains many classes which are inappropriate e.g. disinfectants, dentrifrices and the like. The next question is how valid is it to compare CNS drugs with topical steroids with anti-cancer drugs. The danger is that the analysis will tend to produce the lowest common denominator (like the rule-of-5), rather than a stunning insight into molecular diversity. There is also the issue of reverse sampling: how valid is it to deduce the properties of the universal set of biologically active molecules from a subset? The properties of previous drugs may have been driven mainly by bioavailability, or towards making analogues of the natural substrate. Using this data forces an unnatural conservatism into our diversity models. It is also interesting to reflect on what is meant by activity and inactivity. Any molecule will bind to any receptor, although the affinity may have any value between picomolar and gigamolar. If the binding event is viewed in terms of molecular interactions, then interesting, specific binding can be characterised by affinity constants lower than 1000 nM. However, it is not uncommon to find affinity constants of 1000 nM that are mainly due to solvophobic interactions forcing the ligand to associate with the receptor (particularly for hydrophobic compounds like steroids). At 100 nM, some specific non-covalent interactions are being formed, and at levels below 10 nM, there are at least three of the specific interactions present, according to Arien's hypothesis [13]. It should be clear that the activity is a continuous phenomenon, and that drawing an arbitrary division is a hazardous ploy. Furthermore, whilst one can be fairly sure why a compound is active, it is much harder to be precise about why a compound is inactive. Was it the wrong pharmacophore, a steric bump, poor solubility and so on. This issue is covered in the literature on hypothesis generation. Despite all these caveats, two groups have followed such an approach, and claim to be able to distinguish a potential active from a potential inactive, with reasonable confidence. Such results cannot be ignored, and will be of use in the early phases of library design, where the basic feasibility of the library and the reaction are being considered.

Molecules can be described in many different ways, some of which are closely correlated. How then are the different descriptors to be correlated and combined into an overall description? One solution, suggested from the field of QSAR is to autoscale the descriptors. This at least puts everything on an equal footing. However, one may not want to put an equal emphasis on molecular weight as opposed to the number of pharmacophores expressed. Furthermore, changes in the relative weights of the descriptors will lead to libraries of different composition. This question is as yet unresolved, and we suspect that it may have to be dealt with case by case.

Present day molecular descriptors are incomplete, as they have been devised as the result of a compromise between ease of handling vs rigour. A 2D descriptor based on functional groups (e.g. a Daylight or MACCS key) does not contain much useful information about flexibility, or the relative arrangements of the functional groups. A 3D descriptor, such as a pharmacophore key, contains this information, but sometimes in a crude form. 3D descriptors are hard to formulate properly. In the example cited, the pharmacophore centres need to be carefully defined, as do the distance ranges and the conformational analysis parameters used to generate the key. Even the starting geometry can affect the final description. This issue has been covered extensively in the work of Mason *et al.* on the relative advantages of 3-centre and 4-centre pharmacophore keys [14]. Experiments need to be done to ensure that the similarity principle holds, before using a descriptor in library design, to assess how much its imperfections will affect the design.

1.8 Data handling

Discussions of the huge numbers of compounds that could be produced by combinatorial chemistry often focus on the chemistry or design side. An issue that is often ignored is one of handling all the information that is generated when making, analysing, purifying, storing and testing large numbers of compounds. How does one handle all the data, in a way that adds value to the whole operation. Data *per se* is useless if it cannot be assembled, analysed and organised into a coherent scientific hypothesis. At the present time, this issue is proving to be a major headache for many pharmaceutical companies, and millions of dollars and thousands of man-hours have been spent on trying to devise solutions. The data stored should be more than that required just for book-keeping purposes, to allow the deduction of an SAR and the faster optimisation of a lead series, as set out in the criteria for adding value. Is it worth, for example, storing information on compounds rejected from the library and the reasoning behind this rejection? Can the SAR information gleaned from screening a current library be used in a timely fashion to make the next library faster and better (than a competitor might)?

A good information handling system will allow ideas of good manufacturing practices to be applied to remove bottlenecks in the drug design cycle, so that the technologies of combinatorial chemistry and high-throughput screening can be used to their maximum advantage.

1.9 The role of binding sites in library design

Nature has produced the target sites by a long process of evolution. Although there are strong similarities between binding sites on different proteins for an identical natural ligand, there is some diversity of structure within the binding site and within the molecular architecture holding the site together. This aspect of structural diversity within functionally similar sites has received only superficial attention and little systematic analysis has been applied to the problem with respect to drug design.

Why is it important to recognise the problems posed by molecular diversity within functionally related sites? The answer is simple. Diversity in sites offers the key to specific drug design. Decades of experience from studies of drug-receptor interaction have shown that modifications to the structure of small ligands can reveal a wealth of receptor subtypes; empirical classification systems for receptors evolved before sequence data became available. Now that we have both sequence and structural data about many ligand binding sites, it should be possible to design molecules to have specificity for chosen subtypes of binding sites. Before this goal can be achieved, a great deal of detailed analysis on different ligand binding sites will be necessary to elucidate how evolution has accommodated structural changes in the binding site.

A further question that has to be addressed is how the structure underlying the site has been conserved. One assumes that the way in which the site is built up is the same for each class of binding site although the assumption has never been tested. These complexities in architecture of the site foundation and the site itself are ripe for exploration and will have a major impact for drug design.

Suppose that a binding site contains 20 amino acids that lie adjacent to the ligand. If 10 residues in common are judged to be obligatory for binding, then the remainder may be used to create specificity for different design strategies. Furthermore, suppose that any r residues from n residues available for specificity are required to create binding specificity, then the combinatorial number, $C(n,r)$, of possible design strategies for specificity can be gained from the equation

$$C(n,r) = \frac{n!}{r!(n-r)!} \tag{1}$$

Thus, if $n = 10$ and $r = 5$, there are 252 different design strategies possible. Of course this is a gross oversimplification of the problem; the actual size of the problem is dependant on how many mutations have occurred at each of the n residue positions and how many residues present a realistically different portion of the site for specificity. Even a simple hand-waving exercise as this, illustrates the fact that structural diversity in the site offers an enormous scope for specific design. Furthermore, the design problems to be addressed in seeking specificity from site diversity are very different from normal automated design constraints.

Most natural ligands are flexible molecules and contain key interacting points, ligand points, which lie adjacent to site points in the receptor-binding site. Thus molecular diversity within the site may have evolved to be able to accommodate the same ligand by firstly enabling different conformations of the ligand to bind and secondly, by allowing different ligand points to interact with different site points. In the following discussion these two features are documented. Our aim in this chapter is simply to illustrate the problems for drug design that are presented by site diversity within a set of proteins containing these functionally similar sites.

2. STRATEGIES FOR SITE ANALYSIS

2.1 Choice of a test set of binding sites

We need a suitable group of functionally similar binding sites to study so that any discriminatory features can be identified. Many kinases have a role in cell signalling mechanisms. Large numbers of kinases have been found from gene sequencing in the Human Genome Project. It is widely believed in the pharmaceutical industry that some of the kinases could be key therapeutic targets for drug design. The Brookhaven Protein Data Bank (PDB) contains a mass of good, high-resolution protein structures co-crystallised with nucleotides. The nucleotide binding sites together with their associated ligand structures have not received the comprehensive structural analysis that they deserve. The strategic question that eventually has to be resolved is: to what extent can specific ligands be designed to this plethora of similar sites? Here we sketch out the results of a preliminary survey of ligand binding site data for a number of adenine di- and triphosphates (ADP, ATP) and their structural mimics, co-crystallised with their proteins. The aim is to focus only on the adenosine and its connected two phosphate moieties; ADP is treated as a substructure of ATP.

The ADP moiety is a very flexible structure and could assume a variety of conformations in the binding site. However, in practice there are only a small number of bound conformations that are actually observed. The architecture of the site, conserved by evolution, appears to restrict the conformations found. The approach described in Section 2 is to apply classification methods to the ligand conformations and then to hunt for structural and functional correlations derived from the site which are associated with the observed ligand conformations.

2.2 Alignment of binding sites

If there is a significant sequence difference within a set of sites, the alignment and superposition of sites for comparison becomes non-trivial. A simple strategy would be to superpose the atoms of the backbone and use that superposition as a reference frame. In many respects the backbone is an artefact of the site and the drug designer would like to have a superposition of the surface atoms of the site. However, if there is no obvious correspondence between the atoms of one site and another, comparisons will be dominated by shape similarity and not necessarily by local functionality. An analysis based solely on the sites can lead to problems if their ligands are aligned with different binding modes or different ligand conformations.

Furthermore, binding sites are usually composed of many more atoms than there are in the ligand and thus the comparison based on sites only is potentially more difficult to handle.

An alternative method would be to superpose the ligand structures after they have been divided into conformational classes. The ligand conformation then becomes the reference frame for the superposition of the sites. With the nucleotide binding sites studied here, the ADP moiety conformation provides a reference frame for comparing those sites which have a similar shape affecting the ligand conformation. This procedure allows the shape and functionality of the classified sites to be compared unambiguously. Furthermore, it provides the user with a better foundation for three-dimensional functional motif searching.

2.3 Choice of ligand dataset

There are many high–resolution purine-nucleotide/protein co-crystal structures in the Protein Data Bank. Twenty six complex domains with a resolution < 2.50Å were used: these contained 13 ADP molecules, 10 ATP molecules and 3 mimics. Multimeric complexes were reduced to a single representative example site to avoid biasing the dataset. The sets of complete residues within 4.5Å of the Van der Waals surface of each ligand were defined as the ligand binding sites and their co-ordinates extracted similarly from each complex.

2.4 Analysis of the ligand conformations

The ligand conformations need to be placed in classes. This analysis is based on a geometric disposition of all structurally identical atoms of the ADP moiety. The algorithm of Gerber & Müller [15] was used to superpose all the ADP moieties by pairwise root mean square distances between the pairs of ligands using corresponding atoms; the weighted sum of the mutual least-squared deviation was minimised. Cluster analysis was applied to the resulting pairwise rms difference-distance matrix using a geometric (cluster centre), agglomerative method employing the minimal variance between the individual ligands [16]. The number of significant clusters in a dataset was defined by Mojena's Stopping Rule Number 1 using a level of significance set at $p < 0.05$ to define the number of acceptable clusters [17]. Acceptably different clusters are found on the dendrogram where the significance line cuts the descenders at a scaled fusion distance corresponding to the significance value. Caution must be exercised in interpreting cluster significance levels by this method; it is sometimes possible to take a significant cluster from the total dataset, treat it as an isolated cluster, and subdivide it further into subsidiary significant clusters. This stepwise significance testing on subsets within the data can proceed until no significant difference is found; this procedure identifies significant hierarchically related subsets of clusters.

The cluster data for the ADP moiety conformations for 26 protein crystal complexes is presented in figure 3. It can be seen from the dendrogram that two principal clusters of conformations for the ADP moiety in binding sites are found. 17 molecules are found in class 1 and 9 molecules are in class 2. Class 2 cannot be significantly subdivided. Stepwise significance testing indicates that class 1 can be further sub-divided into sub-class 1a (14 molecules) and sub-class 1b (3 molecules) (figure 4). Sub-class 1a can be divided into two further clusters, with 9 members in sub-cluster 1 and 5 members in sub-cluster 2 (figure 5). Thereafter no further subdivision of these families yields significantly different clusters. After removing sub-class 1b, sub-class 1a, which was originally composed of 2 clusters of 7 members, now has the membership reassigned into 9 and 5 members in the two families. The ADP moieties 1pfk_a and 1rkd have been moved into family 1 as the scaled fusion distance from this pair is now closer to the other members of family 1.

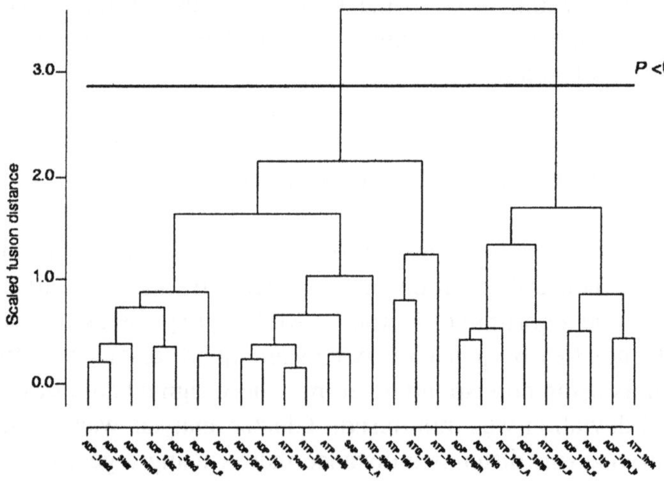

Figure 3. The conformational classes for ADP moieties in protein-ligand co-crystals. Two clusters are significantly different according to the significance line drawn where $p < 0.05$. The cluster issuing from the left descender is termed class 1 ($n = 17$, $mean_{rms} = 11.32$, $S.D._{rms} = 4.43$) and that from the right descender is class 2 ($n = 9$, $mean_{rms} = 6.31$, $S.D._{rms} = 3.43$).

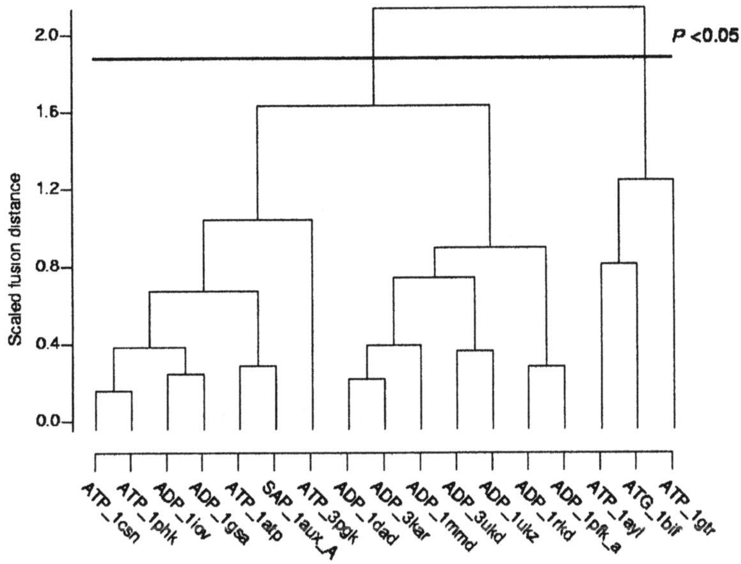

Figure 4. The conformational sub–classes of class 1. Two significantly different sub–classes are found: sub–class 1a (n = 14, mean $_{rms}$ = 7.16, S.D.$_{rms}$ = 3.05) to the left and sub–class 1b (n = 3, mean $_{rms}$ = 2.04, S.D.$_{rms}$ = 2.54) to the right.

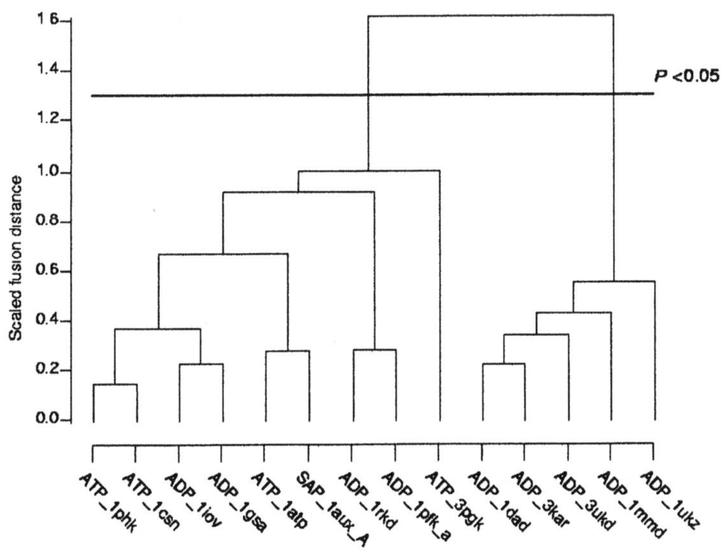

Figure 5. The dendrogram of the subdivision of sub–class 1a into two further families. The ADP moieties 1pfk_a and 1rkd are automatically reassigned, giving two significantly different families: Family 1 (n = 9, mean$_{rms}$ = 3.83, S.D.$_{rms}$ = 2.16) to the left and Family 2 (n = 5, mean$_{rms}$ = 1.52, S.D.$_{rms}$ = 1.24) to the right.

The conformations within a class can then be viewed by taking the conformation of the ADP moiety closest to the cluster centroid to be the reference ligand conformation; the other class members are then superposed onto that molecule. This ligand conformation has the lowest rms difference from the other members of its family and therefore forms the reference frame for comparison of sites associated with that conformational class. The two families within class 1 are shown in the top left panel of figure 6 (class 1, sub–class 1a, family 1) and the top right panel of figure 6 (class 1, sub–class 1a, family 2). Class 2 is shown in the bottom panel of figure. The separate classes and families in figure 6 are distinct from each other, therefore the question can be asked about whether these separate conformational classes of the ligands reflect different structures of the binding sites and, if so, where?

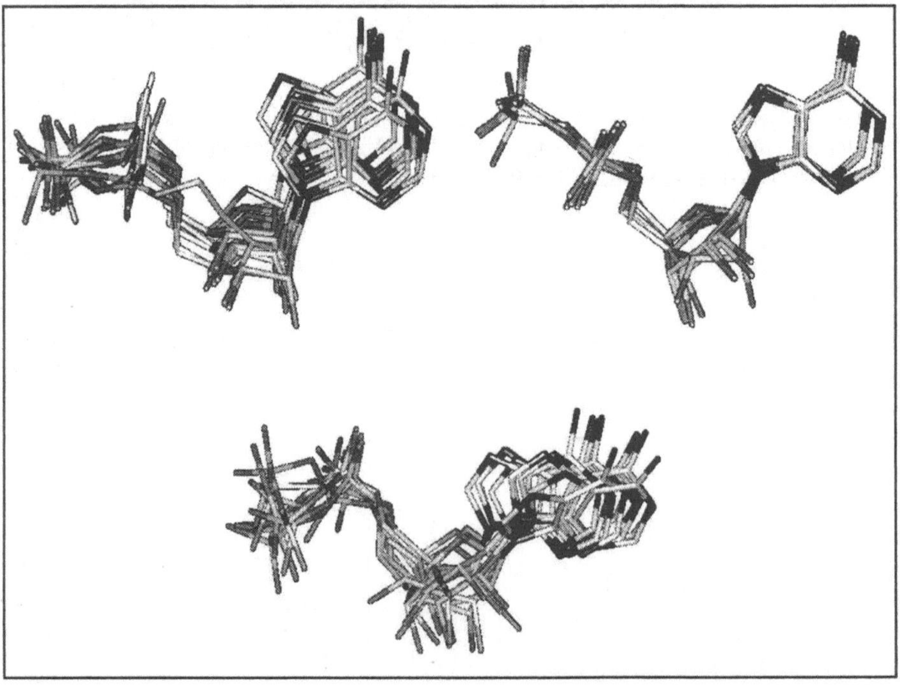

Figure 6. Conformations of the ADP moiety are superposed for sub–class 1a, family 1 (top left) and sub–class 1a, family 2 (top right) and class 2 (bottom).

2.5 Sites corresponding to specific ligand conformational classes

The problem just posed can be investigated by transforming the co-ordinates of each site to the new reference frame from the ligand belonging to the superposed subset. Each ligand conformational class, sub–class or family is then associated with a corresponding set of binding site residues. By definition, these sites are composed of the complete residues within 4.5Å of the ligand van der Waals surface. The superposed sites corresponding to each conformational class, sub-class, or family can therefore be characterised in detail. This gives rise to groups of superposed Cα atom positions indicating regions of similar and dissimilar contact residues.

The preliminary analysis outlined here compares the Cα atom positions in the two families. Figure 7 (top and bottom panel) shows the positions of the Cα atoms of the contact residues. For clarity, the reference ligand is drawn in each case. There is a dramatic difference in the organisation of the site architecture. The top panel has many Cα atoms located round the ribose hydroxyls of the ligand; these atom positions are conspicuously lacking towards the bottom of the picture in the bottom panel. This diversity in what is often assumed to be homologous site architecture has a follow through effect on the distribution of contact residues for the different ligand conformational families. Comparisons of the main types of interactions for both families derived from class 1, sub–class 1a reveals sub–regional diversities. Hydrogen–bonding interactions will be treated here as the developed example, but any property discriminators can be chosen as a subset of the sub–regional diversities.

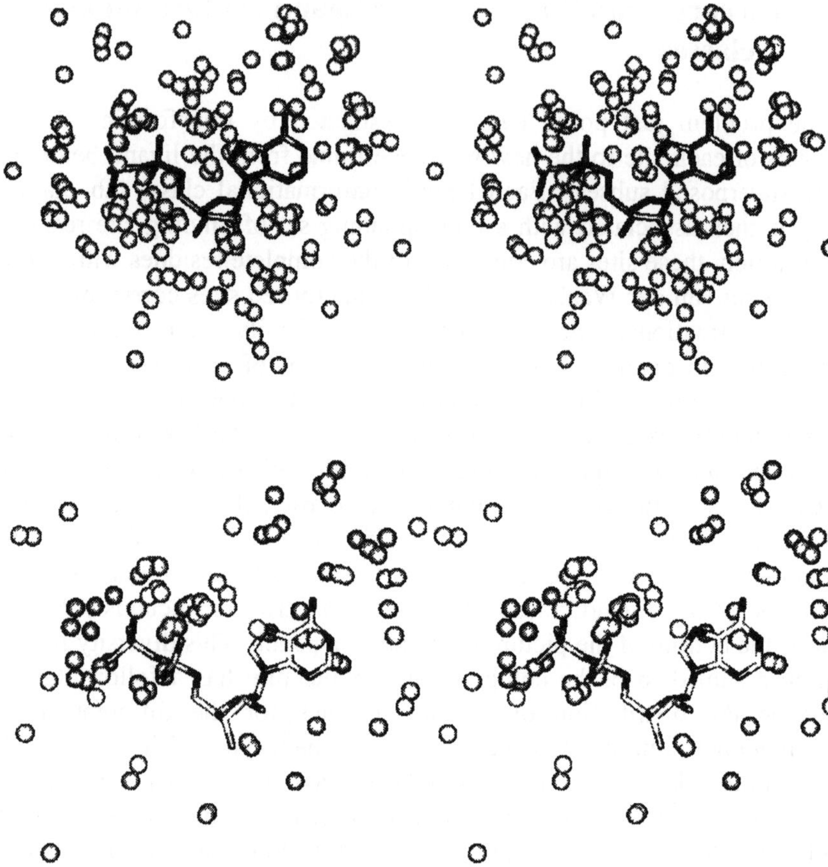

Figure 7. Stereo images of the positions of Cα atoms for the superposed sites in sub–class 1a, family 1 (top panel) and in sub–class 1a, family 2 (bottom panel). Each Cα atom presents a contact residue. The ligands shown are the reference ligands in each case.

2.6 Analysis of Ligand Protein Contacts

Ligand–protein contact (LPC) data describe putative surface interactions between the ligand and the site residues and predict whether they are energetically favourable or unfavourable [18,19]. Putative hydrogen bonds are labelled as backbone, functional group donors or acceptors or amphiprotic. Putative hydrogen bonds with bond lengths less than or equal to 3.5Å are accepted and considered in this analysis.

Corresponding groups of Cα atom positions between the two superposed families of sites identify sub–regions of functionally similar or dissimilar residues. If the hydrogen–bonding sub–regions between the two families interact with the same ligand atoms and share any common modality then they are ignored. In the case of the nucleotide binding–site data presented here, the sub–regions where both families of sites have common interactions with the alpha and beta phosphate oxygens of the ADP moiety are ignored. What remains is therefore informative and table 1 summarises the identified discriminatory hydrogen–bonding interactions (labels in bold) from the important sub–regional diversities for both families (labels a–r).

Table 1. Sub–regional diversities derived from LPC data labelled a–r, between the two families. One or more Cα groups may correspond to those in the other family or may not correspond with any region in the other family (n/c). Labels in bold are discriminatory hydrogen–bonding interactions. The sub–regions j and r are only found in some sites within family 1 but such sub–regions are ignored because they are either areas of either aliphatic or hydrophobic residues or charged residues beyond 3.5Å from the ligand atoms and therefore non–interacting.

	Family 1 (n_1=9)		**Family 2 (n_2=5)**	
Sub-region	**Cα group Identifier**	**Frequency, description (ligand atoms)**	**Cα group Identifier**	**Frequency, description (ligand atoms)**
a	1	1/9, Acceptor (ribose $O_{2'}$)	3	2/5, Donor (adenosine N_6 & N_1)
b	2	No hydrogen bonds	3	2/5, Donor (adenosine N_6 & N_1)
c	4	No hydrogen bonds	6	5/5, Donor (phosphate $O_{\alpha2}$, $O_{\alpha3}$ & O_β)
d	5	3/9, Donor (phosphate O_β)	n/c	
e	6	4/9, Donor (phosphate O_β)	n/c	
f	7	1/9, Donor (phosphate O_β)	n/c	
g	15	6/9, Donor (ribose $O_{2'}$ & $O_{3'}$)	n/c	
h	16	5/9, Amphiprotic (phosphate O_α) & Acceptor (ribose $O_{3'}$)	n/c	
i	17	5/9, Acceptor (phosphate O_α)	n/c	
j	18	No hydrogen bonds	n/c	
k	19	5/9, Acceptor (phosphate $O_{\alpha2}$ & O_β)	17	1/5, Acceptor (ribose $O_{2'}$)
l	21	1/9, Donor (adenine N_6)	5 & 4	No hydrogen bonds
m	22	2/9, Donor & Acceptor (ribose $O_{2'}$& $O_{3'}$)	16	2/5, Donor (adenine N_1 & N_3)
n	23	2/9, Amphiprotic (adenine N_7) & Donor (phosphate $O_{\alpha2}$)	11	5/5, Donor (phosphate $O_{\alpha1}$)
o	24	3/9, Donor (phosphate $O_{\alpha1}$, $O_{\alpha2}$, $O_{\alpha3}$ & O_β)	13	No hydrogen bonds
p	25	3/9, Donor & Acceptor (ribose $O_{2'}$)	18	3/5, Donor (phosphate $O_{\alpha2}$, $O_{\alpha3}$ & O_β)
q	27	2/9, Donor & Acceptor (phosphate O_β)	12	1/5, Acceptor (phosphate $O_{\alpha2}$)
r	29	No hydrogen bonds	n/c	

LPC data also provides the relative contact surface area with the site residues. These values can be used to further prioritise the discriminatory hydrogen–bonding interactions. Since it is also possible to identify the specific proteins that contribute to these interactions, the data can be used to contrast the different sites. Table 2 summarises the combinations of interactions needed to discriminate between pairs of sites from both families. The labels in bold or in italics are subsets of the discriminatory hydrogen–bonding interactions from table 1 and are given a higher priority. The labels in bold in table 2 are considered ideal because they are interactions with maximal contact surface areas. The italicised labels (table 2) are interactions represented in only one of the two families.

Table 2. The combinations of discriminatory hydrogen–bonding interactions needed to contrast pairs of sites from the two families. The labels italicised and in bold are subsets of the discriminatory hydrogen–bonding interactions. The labels in bold have maximal contact areas whereas the italicised labels are only represented in one of the two families. The priority is therefore bold-italic, bold, italic and then normal.

		Family 2				
		3kar	**1dad**	**3ukd**	**1mmd**	**1ukz**
Family 1	**1phk**	*eghi* c	*eghi* ck	**ab** *eghi* c	*eghi* c	**ab** *eghi* c
	1csn	*deh* c	*deh* ck	**b** *deh* c	*deh* c	**b** *deh* c
	1iov	*gio* c	*gio* ck	**b** *gio* cp	*gio* cp	**b** *gio* cp
	1gsa	*gio* c	*gio* ckm	**b** *gio* cp	*gio* cmp	**b** *gio* cp
	1atp	**n** *eghi* c	**n** *eghi* ck	**bn** *eghi* c	**n** *eghi* c	**bn** *eghi* c
	1aux	*dgio* c	*dgio* c	**b** *dgio* c	*dgio* c	**b** *dgio* c
	1rkd	**n** *gh* c	**n** *gh* cq	**bn** *gh* cp	**n** *gh* cp	**bn** *gh* cp
	1pfk	***f l*** *de* c	***f l*** *de* cm	***f l*** **b** *de* c	***f l*** *de* cm	***f l*** **b** *de* c
	3pgk	*gh* c	*gh* cq	**b** *gh* c	*gh* c	**b** *gh* c

Not only is it possible to identify combinations of interactions that discriminate *between* the two families, but it is also possible to discriminate certain members *within* a family. Within family 2, interactions "k" and "q" are unique to the binding site from 1dad. Within family 1, the binding site from 1phk has a unique interaction, labelled "**a**", and the binding site from 1pfk has "***f***" and "***l***" both as unique interactions. Given that interactions "***f***" and "***l***" have the highest priority, they can clearly act as principal foci for 1pfk–specific pharmacophore design. Interaction "***f***" corresponds to a βphosphate oxygen interaction, "*l*" is within the hydrophobic pockets and interacts with N_6 of the adenine system. Each interaction occurs at opposite ends of the binding site.

2.7 Discussion

The results suggest the ease with which it is possible to describe molecular diversities between similar ligand binding sites based on the initial rms classification of the ligands. Sub–regional diversities between sets of superposed sites can be characterised by relating contact residue positions to property information from current on–line biological databases. The automation of this procedure not only provides rapid comparisons between conformational sets of sites but also improves the efficiency of directing *de novo* pharmacophore design by maximising contact information at little computational cost. The sub–categorisation and characterisation of functionally identical binding sites also lends itself to more efficient motif searching and the prediction of a conformationally specific binding site reducing the dependency on homology models of entire structural domains.

In the future it will be necessary to combine diversity of site analysis, for functionally similar regions, with focused diversity methods for small molecules. The methods of molecular diversity outlined in this book could then be combined with diversity procedures so that drug design can be channelled down avenues that lead to site specificity at an early stage in the design process.

3. CONCLUSION

Methods for the analysis of molecular diversity are a powerful tool for drug discovery, when allied to the related technolgies fo combinatorial chemistry and high-throughput screening. However, there is a price to pay, in terms of the many complex logistical and theoretical issues that arise from the size and scale of the operation. These issues have been presented, but require a more through and lengthy discussion than can be provided in this chapter; these topics are covered in sufficient depth in the accompanying papers. A method for the analysis of binding sites, that can be used to explore the common features and the differences between a set of related binding sites, has been presented, based on a survey of ligand binding contacts. The method is applied to the analysis of nucleotide binding sites, and has been shown to highligh the key interactions for specificity and affinity in a rapid and automated fashion.

REFERENCES

1. Floyd, C.D., Lewis, C.N. and Whittaker, M. More leads in the haystack. *Chem. Br.*, 1996, 31-35.
2. Kubinyi, H. Similarity and dissimilarity: a medicinal chemist's view. *Perspect. Drug Disc. Des.* 1998, **9/10/11**, 225-252.
3. Houghten, R.A. Combinatorial Libraries: Finding the needle in the haystack. *Current Biology*, 1994, **4**, 564-567.
4. Newton C.G. Molecular Diversity in Drug Design. Application to High-Speed Synthesis and High-Throughput Screening. In: *Molecular Diversity in Drug Design*, Ed. Dean P.M. and Lewis R.A., Kluwer, 1999, Ch. 2.
5. Barnard, J.M., Downs, G.M., Willett, P., Tyrrell, S.M. and Turner, D.B. Rapid diversity analysis in combinatorial libraries using Markush structure techniques. 213th ACS National Meeting, San Francisco, California, April 13, 1997.
6. Ecker, D.J. and Crooke, S.T. Combinatorial drug discovery: which methods will produce the greatest value? *Biotech.*, 1995, **13**, 351-360.
7. Lajiness, M. Evaluation of the Performance of Dissimilarity Selection Methodology. In *QSAR: Rational Approaches to the Design of Bioactive Compounds,* Eds Silipo, C. and Vittoria, A., Escom, **1991**, pp. 201-204.
8. Patterson, D.E., Cramer, R.D., Ferguson, A.M., Clark, R.D. and Weinberger, L.E. Neighbourhood Behaviour: A Useful Concept for Validation of Molecular Diversity Descriptors. *J. Med. Chem.,* 1996, **39**, 3049-3059.
9. Gillet, V.J., Willett, P. and Bradshaw, J. Identification of biological activity profiles using substructural analysis and genetic algorithms. *J. Chem. Inf. Comput. Sci.*, 1998, **38**, 165-179.
10. Cramer, R.D., Clark, R.D., Patterson, D.E. and Ferguson, A.M. Bioisosterism as a Molecular Diversity Descriptor: Steric Fields of Single Topomeric Conformers. *J. Med. Chem.,* 1996, **39**, 3060-3069.
11. World Drug Index, Derwent Publications Ltd. 14 Great Queen Street, London, WC2B, UK.
12. Daylight Chemical Information Systems, Inc., 27401 Los Altos, 370 Mission Viejo, CA 92691 USA.
13. Farmer, P.S. and Ariëns, E.J. Speculations on the design of non-peptide peptidomimetics. *Trends Pharmacol. Sci.*, 1982, **3**, 362-365.
14. Mason, J.S. and Hermsmeier, M.A. Diversity Assessment. *Curr. Opin. Chem. Biol.* 1999, **3**, 342-349.
15. Gerber, P.R. and Müller, K. Superimposing Several Sets of Atomic Coordinates. *Acta Crystalographr. A*, 1987, **43**, 426–428.
16. Ward, J.H. Hierarchical Grouping for Evaluating Clustering Methods. *J. Am. Stat. Assoc.*, 1963, **58**, 236–244.
17. Mojena, R. Hierarchical grouping methods and stopping rules: An evaluation. *The Computer Journal*, 1977, **20**, 359–363.
18. Sobolev, V., Wade, R. Vriend, G., and Edeman, M. Molecular Docking Using Surface Complementarity. *PROTEINS: Structure, Function and Genetics*, 1996, **25**, 120–129.
19. Sobolev, V. and Edelman, M. Modeling the Quinone–B Binding Site of Photosystem–II Reaction Centre Using Notions of Complementarity and Contact–Surface between Atoms. *PROTEINS: Structure, Function and Genetics*, 1995, **21**, 214–225.

Chapter 2

Molecular Diversity in Drug Design. Application to High-Speed Synthesis and High-Throughput Screening

High-Speed Synthesis and High-Throughput Screening

Christopher G. Newton
Dagenham Research Centre, Rhone-Poulenc Rorer, Rainham Road South, Dagenham, Essex RM10 7XS UK

Key words: Pharmacodynamics, Pharmacokinetics, Bioavailability, Solubility

Abstract: The goal of High Speed Synthesis, High Throughput Screening and Molecular Diversity technologies in the pharmaceutical industry is to reduce the cost of finding good quality leads against a pharmaceutical target. Good quality leads should allow faster optimisation to candidate drugs. It is vital to maintain this perspective when discussing the advantages of these enabling technologies and the large costs associated with their implementation and running. The focus of this paper will be on reviewing the factors that seems to explain why some compounds are better leads and candidate drugs than others. This will help to set out the strategic decisions that have to be made, to try to optimise the benefits and synergies of HSS, HTS and diversity. The conclusion is that considerations of pharmacological conformity - that the molecules designed have the best chance of being fit-for-purpose - should be placed before considerations of how diverse molecules are from one another.

1. INTRODUCTION

Why should the pharmaceutical research manager be interested in the concept of diversity when optimising the discovery of new pharmaceuticals? It is because the cost of finding leads against a pharmaceutical target, which can then be optimised quickly to a candidate drug has become very expensive, and it has been claimed [1] that considerations of molecular diversity in selecting compounds for screening may reduce this expense.

This chapter seeks to place the diversity concept into perspective when considering the research process.

The cost of importing the new technologies of high-speed synthesis and high-throughput screening into pharmaceutical research for lead finding is phenomenal. The new technologies of high-speed synthesis require robotic systems, few of which in 1998 retail for under £100,000, and few of which are yet capable of synthesising more than a couple of hundred compounds at one time (ACT 496: £130,000 96-at-once, Argonaut Nautilus: £130,000, 24-at-once, Bohdan: £150,000, 96-at-once). Most syntheses require dedicated synthesis of building blocks, and development chemistry before transfer to robotic apparatus. Robotic synthesis, which can take many hours, is then followed by isolation and analysis steps, often off-deck, which require several days to perform. Registration, storage and submission of millions of samples equally requires time and expensive equipment (the Haystack™ system marketed by the Technology Partnership in the UK costs several million pounds per unit). On top of the capital and maintenance costs, the material costs of making compounds, even on the milligram sale is not insignificant. Costs of libraries from 3rd party suppliers are currently in the range of £10 to £100 per compound for non-proprietary compounds, £100-£500 for novel, proprietary compounds.

The available screening systems of 1998 tend to be higher capacity than the synthesis systems, but are equally expensive. The Zymark Allegro™ system operational at the RPR laboratories in Collegeville, USA is capable of screening some 100,000 assay points per day, in fully automated mode, total capital expenditure for the system, enclosed ventilated cabinets and laboratory was over $1M. Similarly, the revenue expenditure in screening can be between £0.1 to £2.00 per point, depending on the cost of reagents. Hit rates in screening are generally low for many targets (0.1% has been quoted as typical when screening historical, large corporate collections), but nonetheless, this equates to 100 hits per 100,000 compounds screened. Few companies have the capability of optimising more than a few lead series to development candidates. Furthermore, to the chagrin of their biological colleagues, many hits are often rejected as leads by chemists due to their intractability. The conclusion of this analysis is that hits, when found, should be optimisable.

The costs of registering all the data associated with the synthesis, storage, screening and of analysing the compounds must not be discounted. Most companies are wrestling with various combinations of relational databases, chemical registration packages and efficient search engines, to improve the decision making processes, which become daunting when operating on such vast throughputs of compounds and screening points.

Given the costs of synthesising, analysing and screening hundreds of thousands of compounds to get two or three worthy lead series, it is hardly surprising that attention has turned to the design of better screening sets. The promise is that such design will produce the same number of high-quality lead series whilst screening far fewer compounds, and ensuring that every hit compound obtained, is fit-for-purpose as a lead. Diversity is one method claimed to fulfil this promise. However other factors must be considered before molecular diversity factors are built into library/compound design.

Much attention has been given in conferences over the past few years to the concept of diversity of screening sets, (whether within the set, or between sets), as a method of reducing the costs of lead finding and subsequent optimisation times. Thus, the diversity of compounds may be considered to be a universe truly as unlimited as the number of individual compounds that can be prepared, which probably exceeds the number of carbon atoms on the planet. However, within that diversity, degrees of similarity between molecules has long been a useful method of partitioning the infinity of molecular space, whether the partitioning be used as a guide to molecules of similar properties, comparable use, common patentability, or merely as a convenience in writing reviews. However, diversity alone is insufficient as a method of optimising the drug discovery process. Although it is important, it is subordinate to a greater paradigm, the need for molecules to be made which are "drug-like", molecules possessing the required "pharmacological conformity". Similar requirements for drug-likeness have been expounded by Mitscher in a recent review [2].

2. CONSIDERATION OF PHARMACOLOGICAL CONFORMITY BEFORE MOLECULAR DIVERSITY.

The contention is that the design criterion of pharmacological conformity must be satisfied first, in any corporate collection or combinatorial library, before consideration of molecular diversity is given. The maxim is that every compound should be immediately viable as a lead if it meets the potency criteria in primary screening; all other features making compounds unworthy as commercial drugs should be designed out before synthesis. Clearly, any lead that already has the "drug-like" qualities (that have long been discussed in qualitative fashion, but are only now being analysed in quantitative manner), will imply a reduced optimisation time, especially if the general principles of bio-availability, lack of general toxicity, stability, solubility and crystallinity have been considered in the compounds constituting the lead series. Design quality in the submitted compound sets will reduce the need to perform a decision making process on leads, and lengthy multi-parametric parallel optimisation problems could be avoided.

Consideration of the sub-class of molecules that may be considered as drugs, automatically places a boundary to the sorts of molecules to which considerations of drug diversity should be applied. The boundaries that enclose the organic molecules which are drugs, or candidate drugs, (often termed "drug-like" molecules) have long been recognised, although Messer's [3] search for molecules with a "biological look" was not tightly defined. Recently, however, consideration as to the properties by which the drug class of organic molecules may be defined, has been presented. An appreciation of the property boundaries, and the rationale for consideration in drug design should be undertaken, before consideration is given to the partitioning of the molecules that lie within the boundaries. It should also be appreciated that the boundaries of the galaxy of drugs and potential drugs from the rest of the universe of molecules is diffuse, likely to change with the advent of new discoveries, and may be redrawn, according to the growth of experience. It follows that the arguments and definitions given below concerning pharmacological conformity mirrors the experiences of the author and of the present day.

A molecule requires three types of general property to be acceptable before it can be a drug. These general properties have been termed pharmacodynamic, pharmacokinetic, and pharmaceutic [4]. When creating sets of compounds, be they large corporate collections or combinatorial libraries for high-throughput screening, it will be to the great advantage of the medicinal chemist if the lead molecule generated already contains within it the general attributes of a drug. Indeed, it may be regarded as the absolute responsibility of the CADD expert and the medicinal chemist engaged in high-speed synthesis, or of corporate collection assembly, to ensure that what enters screening, has all such attributes. By judicious building-in of pharmacological conformity into such a screening set, the downstream activities of lead optimisation should be shortened.

2.1 Pharmacodynamic Conformity

Molecules that elicit pharmacological responses may be small and contain no carbon atoms (nitric oxide), or be large organic proteins (β-interferon). They may form covalent bonds with their targets (aspirin, carmustine) or non-covalent bonds with their targets (lovastatin). They may contain a large number of heteroatoms in comparison to the carbon count (DTIC) or very few (progesterone). However, the general features of molecules that usefully interact with proteins are known [5], and the interaction types have been classified. Thus, drugs may interact with their targets through charge pairing, hydrogen bond donor-acceptor, or acceptor-donor non-covalent bonds, by possessing centres of hydrophobicity that interact with similar domains on target proteins, and by π-bonding interactions (e.g. aromatic edge or face stacking, or stacking of amide bonds over aromatic rings). In general, three such interactions are regarded [3] as essential for useful, discriminatory binding, and molecules with two or fewer interacting groups are usually disregarded as not fulfilling the requirements of pharmacodynamic conformity.

Molecules which are flexible may display many pharmacophores (combination of pharmacophoric groups) depending on the molecular conformations that are accessible at physiological temperature. Very many conformations can lead to the display of many thousands of pharmacophores. This might be considered an advantage, in that an ability to display many conformers might imply a greater chance of a hit in screening, however the number of molecules populating each pharmacophore must also be considered, since the concentration apparently available of each pharmacophore of the molecule will affect the apparent potency of the molecule. Notwithstanding how the measure of inherent flexibility is defined (flexibility index, number of freely rotatable bonds, number of displayed pharmacophores per individual molecule), literature analyses of databases of medicinal compounds clearly show that there is a reasonable upper limit to the flexibilty of drug-like molecules. This has been demonstrated in an analysis of datasets presented by Lewis, Mason and McLay [6] and in this case shows that an upper bound of 8 in the MOLCONN-X derived flexibility index seems reasonable.

A second degree of pharmacodynamic exclusion concerns molecules containing reactive groups. The current paradigm is to reject reactive molecules as potential drugs, defining "reactive molecules" as those which form irreversible links to proteins (with the exception of drugs destined as cytostatic agents in cancer, or some anti-bacterial agents). In general, medicinal chemists will actively remove reactive molecules from lists of designed molecules for synthetic preparation, remove them from selection of screening sets from corporate collections, and in particular, ensure that such molecules are not present as impurities in combinatorial libraries. There are many examples in the literature, where apparent activity in a biochemical or pharmacological screen is due to reactivity in a chemical series. Hence, a filter: "reactive filters" [6] should be applied to a potential screening set, to remove compounds that have a significant possibility of forming non-specific covalent bonds to proteinaceous material, or could in some other way achieve a false positive in many screens. A list of such groups appears in table 1, although some groups are borderline and may indeed be found in some inhibitors or receptor antagonists. Table 1, is illustrative, rather than definitive or exhaustive.

Table 3. Reactive filters to improve screening sets

TYPE	EXAMPLES
Active halogens	primary alkyl halides, acyl halides alpha-halo ketones, alpha-halo-ethers acetylenic halides: 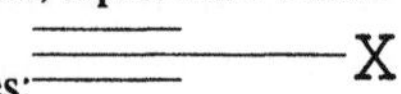
3-membered rings	 Z= O, N, S
Anhydrides	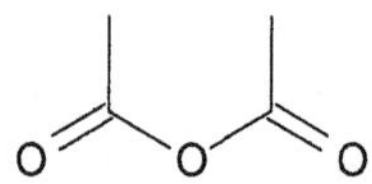
Thiocyanates, cyanates, peroxides, other unstable bond types	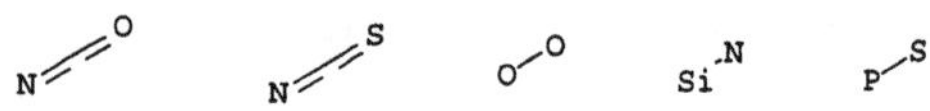 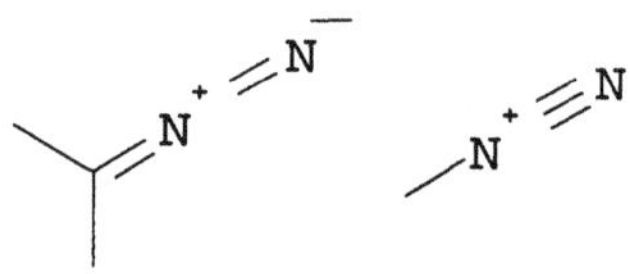
Sulphuryl, phosphoryl, silyl, nitrogen halides	
Sulphonates, sulphinates, silicates	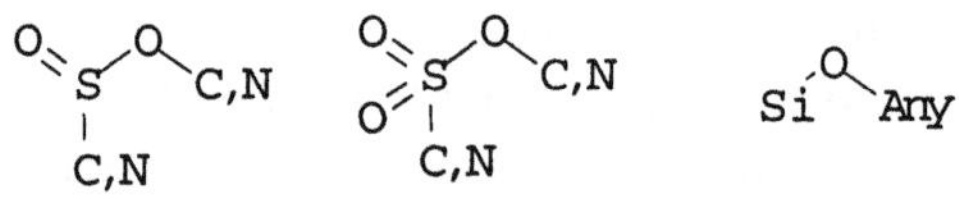
Nitrogen-oxygen systems acyclic nitroamino nitrates nitrones aliphatic-N-oxides nitro groups (limit to 4 per molecule)	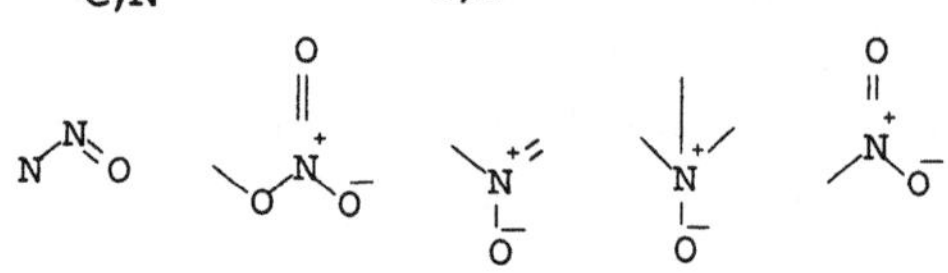
acyclic aminal	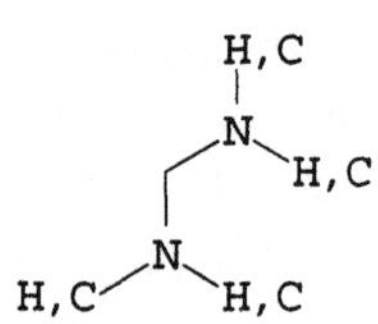
acyclic cyanohydrin	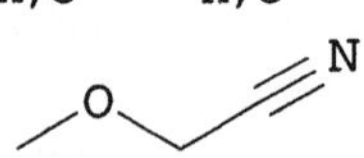
Unstabilised acyclic enols, enolates	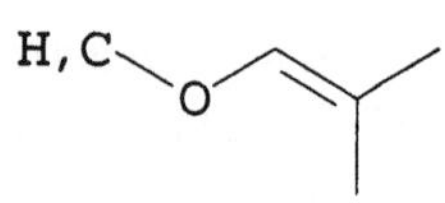

Reactive Michael acceptors - reactivity may be defined by each chemist based upon personal preference.	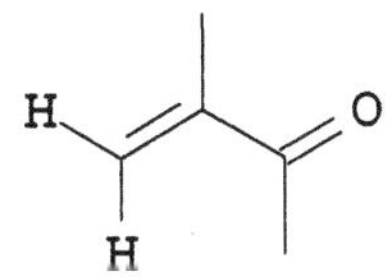
Specific atom-containing compounds	Be, B, Al, Ti, Cr, Mn, Fe, Co, Ni, Cu, Pd, Ag, Sn, Pt, Au, Hg, Pt, Bi, As, Sb, Gd, Se
Hydrocarbons	Any compound not containing at least one O, N, S
Labile esters	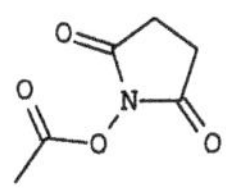 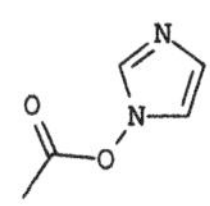 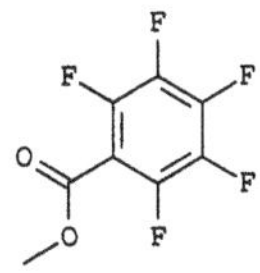
Reactive sulphur compounds (some companies plate mercaptans separately for metalloprotease screens, but reject them for general screening)	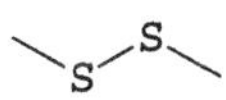 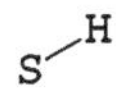 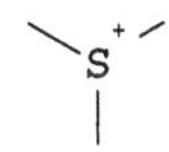
Hydrazines	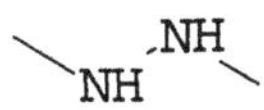
Compounds with very long polymethylene chains (>10) (excepting drugs destined as topical agents).	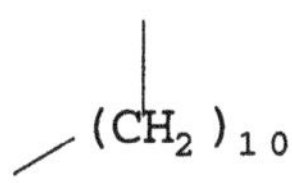

Another class of molecule that is usually removed from lists of molecules to screen, are molecules which are often termed "promiscuous" in their biochemical action, i.e. they tend to have an undesired pharmacological action [6] or a non-selective action over many classes of target. This is often an experience-based exercise, that is subjective to the individual scientist or organisation. In general, molecules of the following classes are removed from design consideration, or from corporate screening sets: steroids, eicosanoids, 3-acylaminobetalactams.

Clearly, pharmacodynamic acceptability boundaries are diffuse, but with consideration of the target, a set of useful boundary parameters may be set within which molecules may be considered "drug like".

2.2 Pharmacokinetic Conformity

The properties of molecules that permit them to be transported to the site of action, i.e. pharmacokinetic conformity, have been published recently [7]. For the majority of drugs which are required to be orally bioavailable, boundary properties have emerged that can be usefully used to delineate "orally deliverable drug-like" molecules.

Properties known to be important in enabling molecules to pass cell membranes, whilst retaining the ability to be transported in plasma, are molecular weight, ionisation constants, lipophilicity, polar surface area, and number of hydrogen donors or acceptors. Lipinski has formulated some general guidelines for boundary definition, which have become known as the "rule of 5" [7]. These are:

1. an upper molecular weight cut-off of 500.
2. a maximum number of hydrogen bond donors in the molecule not to exceed 5.
3. a maximum number of hydrogen bond acceptors in the molecule not to exceed 10.
4. an upper LogP (P = partition coefficient between octanol and water) of 5.
5. rules 1 to 4 only apply to passive transport.

These definitions are an attempt to recognise that penetration through cell membranes is accomplished only rarely by molecules of high molecular weight, unless there is an active transport mechanism. Figure 1 illustrates the consideration [8] of the profile of two data sets, one taken of the Standard Drug File (23747 entries), and the second from a Pharmaprojects data set (5199 entries) shows that "drug-like" molecules have a mean molecular weight of around 350, with a platykurtic distribution around this. In a similar profiling of commercial drugs, Lipinski found that only 11% of drugs had a molecular weight greater than 500 [7].

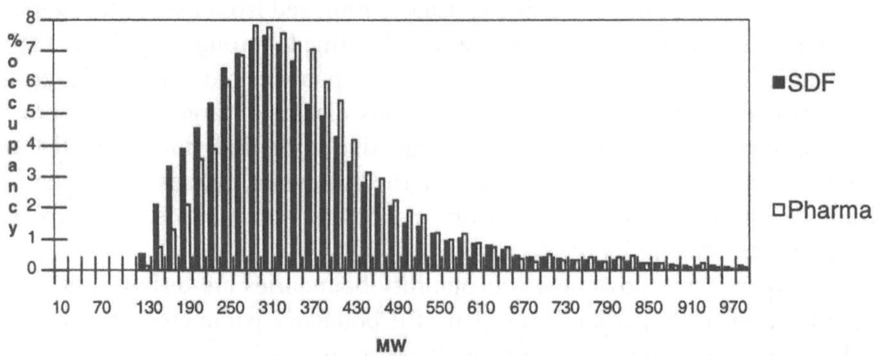

Figure 1. Plot of % compounds falling within Molecular Weight (MW) bin size for an SDF derived dataset (23747 compounds) and a Pharmaprojects derived dataset (5199 compounds)

Lipinski's upper bounds on the numbers of hydrogen bond donors and acceptors can be related to the ability of a molecule to penetrate lipid membranes. Molecules which have groups which can form hydrogen bonds, bind to water; free energy will be required to displace the water as the molecule is transported across the membrane. This energy expensive process can become limiting if too many water molecules have to be removed. Only in special cases, such as that of cyclosporin [9] does internal compensation of hydrogen bonds appear to override the rule as defined by Lipinski.

High LogP values imply high solubility in fat (and good penetration of lipid membranes), but by implication, low solubility in aqueous phases, and hence an inability for the molecule to be transported through the body. Molecules with high LogP values also tend to be the substrates of the metabolising cytochrome P450 enzymes in the liver, in which case, first pass effects can remove much of the administered drug candidate before it can reach its target organ.

Consideration (figure 2) of both a Standard Drug File (24082 entries) and a Pharmaprojects dataset (5279 entries) shows that "drug-like" molecules have a normal distribution, with a modal value of around 2.5, and a LogP of 5.0 is indeed a reasonable upper bound for a candidate drug molecule.

Lipinski states that experimentally, a lower (hydrophilic) boundary to LogP would be expected, for absorption and penetration, but that operationally such a lower limit is ignored because of errors in calculation, and also because excessively hydrophilic compounds are not a problem in laboratories! If a lower bound to LogP is deemed to be appropriate, perhaps reflecting a need for a molecule to have some affinity for lipid, then inspection of the distributions in figure 2 indicate that a lower limit of pharmacokinetic conformity could be set around a LogP between 0 and -1.

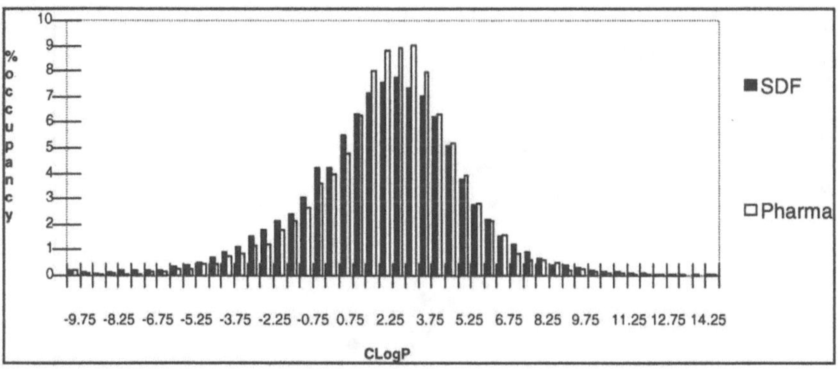

Figure 2. Plot of % compounds falling within CLogP bin size for a SDF-derived dataset (24082 compounds) and an Pharmaprojects-derived dataset (5279) compounds).

Molecules which are permanently ionised (e.g. quaternary ammonium salts) can be quite acceptable drugs in special circumstances, such as drugs which are to be administered i.v., but for drugs aimed to be delivered orally, such permanent ionisation will confer poor pharmacokinetic properties. Hence knowledge of the ionisation pattern (pKa constants) will be required. Monoacids and monobases with pKa values in the range of 3-10 would be a reasonable cut-off for avoiding oral bioavailability problems. Combined with LogP measurements, to calculate LogD (distribution constant of a molecule between octanol and water at pH 7.4) a composite pharmacokinetic criterion can be obtained, with LogD ranging from -2 to 5.

The relationship between the polar surface area of a candidate drug and its oral bioavailability, has been the object of recent papers published by groups in Uppsala [10]. Polar surface area will inversely correlate with lipid penetration ability; a molecule with high polar surface area attracts large numbers of hydrogen-bonded water molecules, requiring input of considerable free energy to displace them before passage through cell membranes can be accomplished. From the data presented in the paper (see figure 3), Palm [10] draws the conclusion that molecules having a polar surface area of >140 Å^2 are less than 10% absorbed. A polar surface area of 120-140 Å^2 thus sets an upper limit for PSA, in the design of oral drugs. It should be noted that polar surface area has a degree of conformational dependence, and calculation can become CPU intensive if a molecule displays many different, energetically available conformers. Nonetheless, calculations made from single low-energy conformations appear to give equally good correlations [11].

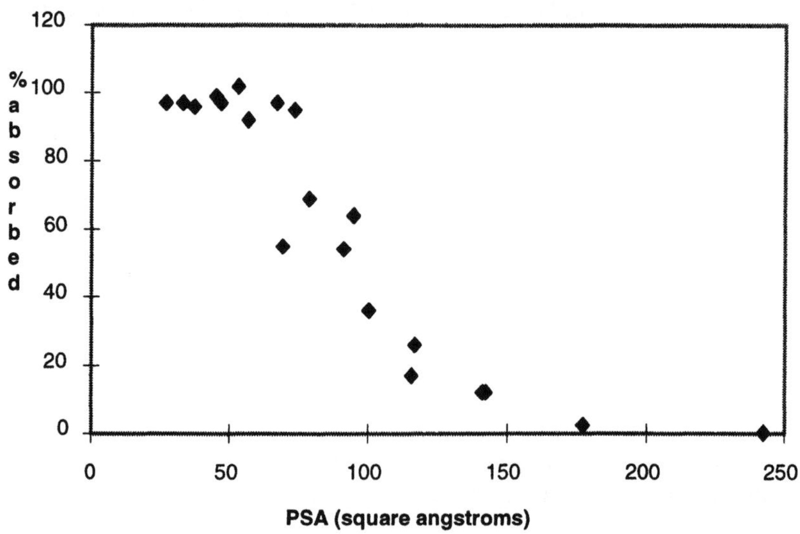

Figure 3. Plot of Polar Dynamic Surface Area (square angstroms) and fractionally absorbed dose in humans, for 20 selected compounds (data from Palm *et al* [10]).

Molecules which are intended to be active in the brain require in addition an ability to cross the blood-brain barrier. The value ascribed to this ability, BB, is defined [12] as:

$$BB = \text{concentration in brain/concentration in blood.} \qquad (1)$$

Since it has been established [13] that this parameter, useful for predicting the pharmacological utility of a compound destined for a CNS application, correlates poorly with the water/octanol partition coefficient above, both CADD scientists and physical chemists have been struggling to establish methods for predicting this useful parameter. Abrahams [12] has defined the following equation as predictive of measured BB values, enabling one to design the "correct" values into a drug molecule, before synthesis.

$$\log BB = 0.04 + 0.20R_2 - 0.69\pi^H_2 - 0.72\Sigma\alpha^H_2 - 0.70\Sigma\beta^H_2 + 1.00Vx \qquad (2)$$

where R_2 is an excess molar refraction, π^H_2 is the solute dipolarity/polarisability, $\Sigma\alpha^H_2$ is the summation hydrogen bond acidity, $\Sigma\beta^H_2$ is the summation hydrogen bond basicity and Vx is McGowan's characteristic volume.

According to Abrahams, this equation is intuitively correct, as it shows the factors that influence blood-brain distribution. Thus, solute size increases BB, whilst polarisability, hydrogen bond acidity and hydrogen bond basicity reduce it.

More recently, a group at Pfizer [14] have also attempted to calculate from the structure of a molecule, its ability to cross the blood-brain barrier, based upon the solvation free energies of the solute in the two immiscible phases. For a range of 63 compounds with a LogBB ranging from -2.15 to +1.04, computed logBB values based upon calculations of the free energy of solvation of each solute in water and in *n*-hexadecane correlated well with experimental values, according to the equation

$$\log BB = 0.054\Delta G^\circ_W + 0.43 \qquad (3)$$

Recent work by Clark has demonstrated the utility of calculated polar surface area as a predictor for blood-brain barrier penetration [15]:

$$\log BB = -0.0148PSA + 0.152ClogP + 0.139 \qquad (4)$$

where PSA is polar surface area. Clearly this gives the experimentalist a calculable value of Log BB to aim for in the experimental design of CNS active drugs.

Thus, when searching for the pharmacokinetic boundaries to drug-like space, many parameters can be considered, many overlap in their descriptive properties, and some are easier to calculate or measure than others. In general, boundary definitions of the useful "drug universe" should only be made upon measured data, but the CADD specialist, called upon by the medicinal chemist, will be required to calculate such properties in advance of molecules being prepared, and indeed, consideration of such properties may be considered as important in drug design as any anticipated fit with a receptor protein or enzyme. Calculation of molecular weight and hydrogen-bond donor/acceptor patterns are trivial. Reasonable quality commercial programmes now exist to calculate LogP, pKa, LogD, but not yet polar surface area and LogBB.

Calculations for pharmacokinetic conformity can be performed for individual libraries, but displays of sheets of numbers becomes tedious for chemists or CADD specialists to analyse. For library design, graphical output of data is preferred. Figures 6 and 7 show graphical data for an Ugi library (figure 4), showing a poor fit relative to the standard SDF database, whereas figures 8 and 9 show an acceptable distribution for a pyrazole library (figure 5). When one wants to improve conformity, direct links from library input files (building blocks) enable library properties to be varied in real time, as the building blocks are changed.

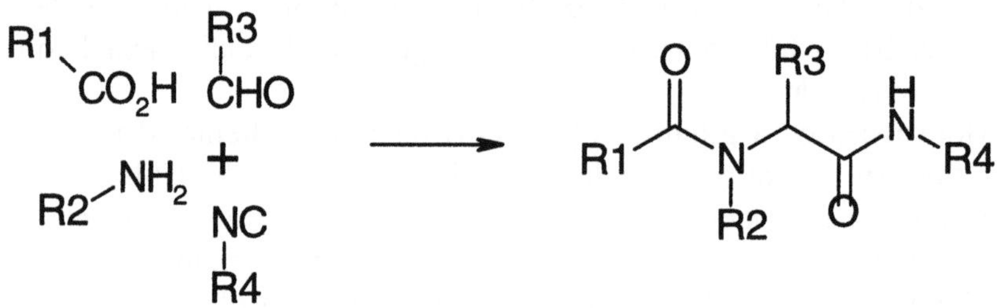

Figure 4. Construction of a library using an Ugi four component reaction

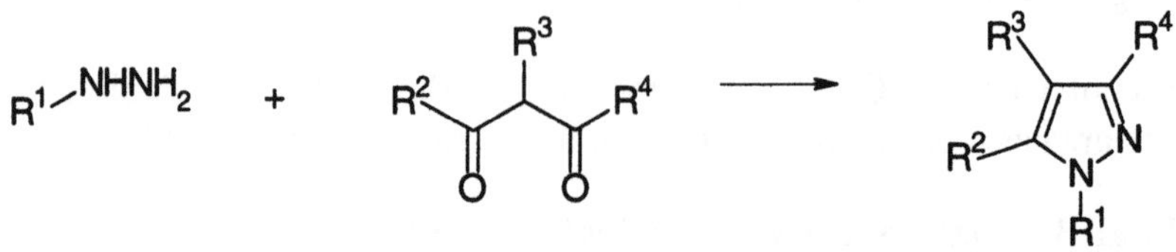

Figure 5. Construction of a pyrazole library yielding four points of diversity

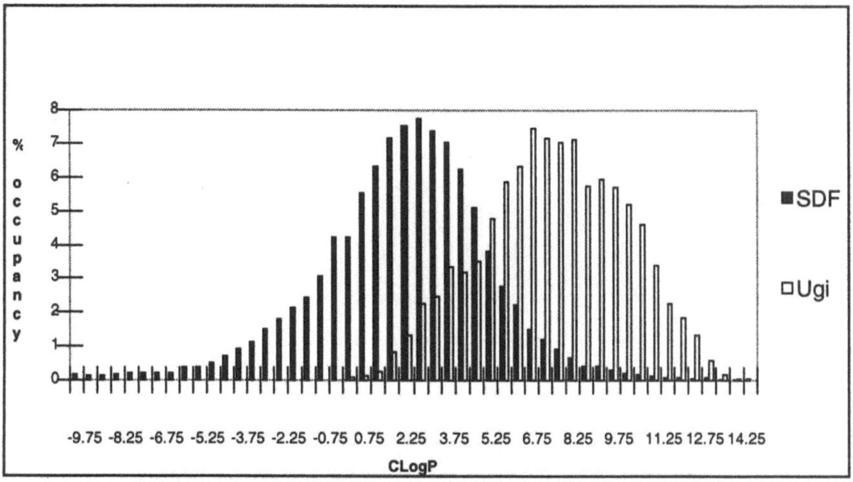

Figure 6. Plot of % compounds falling within ClogP bin size for an Ugi library (3746 compounds) versus that of the SDF-derived dataset (24082 compounds)

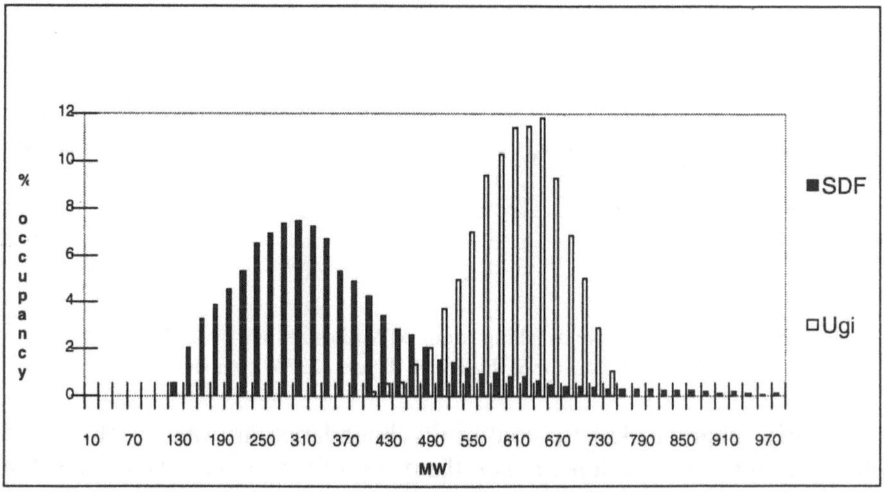

Figure 7. Plot of % compounds falling within MW bin size for an Ugi library (3744 compounds) versus that of the SDF-derived set (23747 compounds)

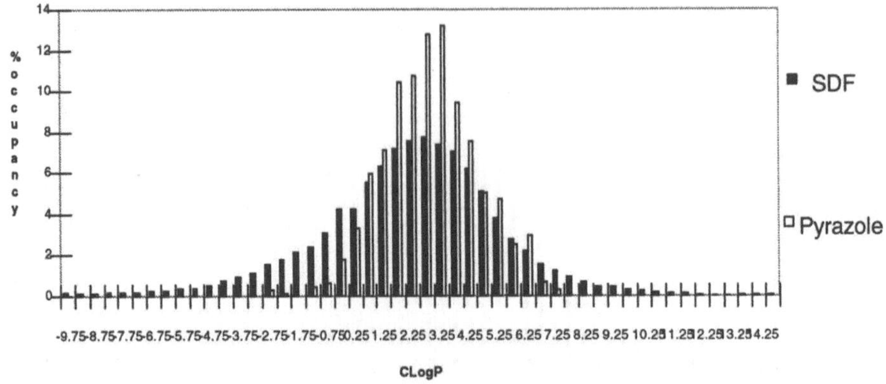

Figure 8. Plot of % compounds falling within CLogP bin size for a Pyrazole library (994 compounds) versus the SDF-derived dataset (24082 compounds)

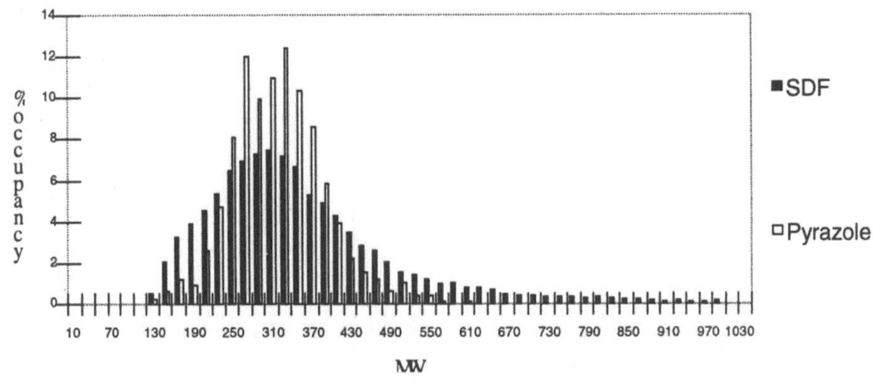

Figure 9. Plot of % compounds falling within MW bin size for a pyrazole library (992 compounds) versus the SDF-derived dataset (23747 compounds).

Clearly, pharmacokinetic conformity boundaries are not absolute, and library design can be tailored to suit the needs of the project. Designing and making molecules or building libraries outside the "normal" pharmacokinetic grounds may well be judged acceptable, and indeed data on such molecules may well encourage a rethinking of the pharmacokinetic boundary criteria, in the future.

2.3 Pharmaceutical Conformity

Pharmaceutical criteria are important for drugs, and such properties include melting points, aqueous solubility, crystallinity, polymorphism and chemical and physical stability. These require optimisation during the drug discovery process.

Melting points of over 100°C are preferred for operational pharmaceutical reasons. Where drugs require pharmaceutical finishing, and manipulation into formulations, high melting point crystals are less likely to deform or melt, during such processing. However, an upper bound to melting point is also envisageable, perhaps around 200°C; higher melting points implying high crystallisation energies that in turn suggest low solubilities. Melting point is an indication of the energy that will be required to break down the crystal lattice; a low crystal lattice energy implies less energy needs to be expended in dissolution. Low melting point is usually an indicator of low lattice energy.

Solubilities of organic molecules in water greater than 1 mg/mL are preferred, as this will enable good concentrations of the molecules at the points of absorption [16]. Unfortunately for the CADD expert who is supporting a medicinal chemistry design programme, the *de novo* calculation of solubility is difficult, although the work of Yalkowski [17], is now being followed up by new research from Huuskonen [18] and Mitchell [19].

Chemical stability (for isolation and formulation) is another pharmaceutical requisite. Fortunately exclusion of many reactive compounds because of the needs of pharmacodynamic conformity will have already excluded many problem classes of such molecules. Various light, acid or base sensitive groups can also be chosen to be excluded from the set of drug-like-molecules. Such sets are usually experience-based, and also are likely to differ from company to company, based upon that experience. Where *de novo* drug design programs are being used to design templates and libraries, it is essential that some rules of chemical sense are incorporated into the output.

Physical stability, the propensity of compounds to adopt different polymorphs depending on methods of isolation is also now beginning to be tackled by the computational chemist. Ideally, molecules which can only crystallise as a single polymorph would be preferred, although preparation of a compound in the desired (most stable) polymorph, can usually be mastered by the experimental chemist. Nonetheless, identification of the 50% of organic molecules likely to crystallise in different polymorphs would be useful, especially if this can be predicted from the molecular structure of the compounds even before they are synthesised.

3. DIVERSITY IN THE CONTEXT OF HSS-HTS

Having delineated the concept of the need for molecules to have "drug-like" properties, and broken this down into boundaries of pharmacodynamic, pharmacokinetic and pharmaceutical properties, the CADD expert and medicinal chemistry colleague can establish a parameterised multi-dimensional boundary to drug like space (the galaxy within the universe of all molecules). Just as a celestial galaxy contains an immense number of individual stars, so within these rather fluid boundaries, lie untold millions of drug molecules. It is beyond the resource capability to make all of these for screening, and certainly far beyond the resource capability to analyse them all. A judicious sub-selection from the galaxy of drug-like properties is still required.

3.1 Diversity in Collections:

Many companies have corporate collections of hundreds of thousands of individual molecules, many will have millions of molecules in the years to come. Many will seek to screen without sub-selection, but other pharmaceutical houses will elect to make sub-selections for cost-effective purposes.

Two types of sub-selections are usual: those based upon a maximal diversity element, and those based upon a minimal diversity element i.e. traditional 2D/3D searching (always maintaining the selected compounds within the constraints of the above pharmacodynamic, pharmacokinetic and pharmaceutical properties).

3.2 Assembly of sets of drug-like molecules containing a maximum diversity element

One of the simplest diversity measurements is to take the established properties of a molecular set (LogP or molecular weight for example), subdivide that property, and then to attempt to populate that space with molecules from the total collection. Partitioning approaches have recently been reviewed by Mason and Pickett [20]. Populations can be equal in density across the property space, or, perhaps more realistically, weighted to reflect known drug population. Where a particular effect is desired, it is also possible to design to incorporate a particular preferment. Such a diversity approach, (looking solely at the properties of molecules) may be independent of any structural information about the molecules, but in a more powerful guise, can take 3D structural information into consideration, e.g. considerations of pharmacophores and their partitioning into bins.

Clustering molecules based upon structural information can differ enormously in the complexity of the programmes used to assemble them. In a simple, one-dimensional structural approach, a bitmap of each structure is created, with bits turned to 1 if the particular structural feature is present, and to zero if not. Comparison of the bitmaps can then be made, and diversity maximised using an index like that of Tanimoto. More exotic approaches involve the calculation of and recording of every three-point pharmacophore present in a molecule, and then an analysis of similarity across the dataset of molecules [6]. Having established a similarity index between molecules within the library (or perhaps between two libraries), an evaluation of how change in library design changes the similarity can be performed. Intuitively, the medicinal chemist is more attuned to pharmacophore analysis, but there is no evidence that the complexity of the analysis has any bearing on the relative utility of the libraries created, other than in satisfying Senior Management that added value to what existed before, may have been created.

The creation of new libraries with planned diversity can be undertaken in two ways. By the very nature of the beast, a 10x10x10x10 four component library, using just the 40 reagents, will generate 10,000 products. Clearly the temptation is to perform the diversity calculation across the reagents, without consideration as to the chemistry that is putting the reagents together, and thus the molecular architecture created. Analysis of the reagent diversity, particularly of each independent reagent without consideration of the others, can lead to a nonsense in the apparent diversity of the products. Just such a case may be envisaged simply: Consider a 2x2 matrix of reagents used to build four thiazoles (Figure 10). Taken separately, the two thioamides are diverse (methyl versus phenyl), and so -are the two α-bromo acids (methyl versus phenyl). However, inspections of the four products that can be made combinatorially from these four reagents show that one pair is similar in structure, because of the nature of the chemistry that has joined the two reagents together, and the pharmacophoric displays are hence essentially identical. Thus, the desire to calculate the diversity of the **products** of a library, which is a much more computationally challenging objective, is required.

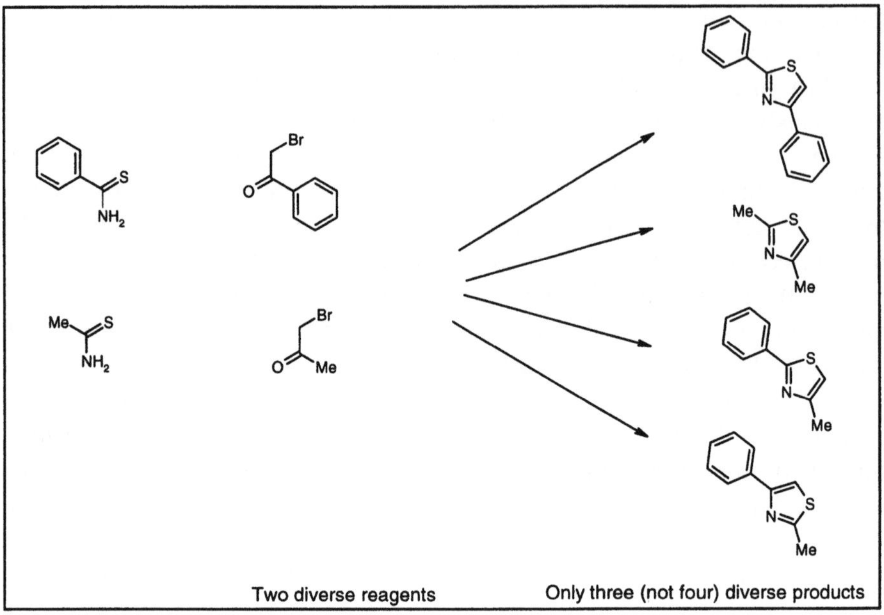

Figure 10. Thiazoles

3.3 Assembly of sets of drug-like molecules containing a minimal structural conformity element.

Many targets submitted to HTS are actually members of large families of related proteins, that experience has shown bind to particular pharmacophoric types. Such privileged chemical groups include metal binding groups such as hydroxamic acids for metalloproteases, monoacids or monobases for G-protein coupled receptors, basic groups such as amidines, for serine proteases. Many medicinal chemistry practitioners prefer to begin screening against a new member of a particular class of target with collections or designed libraries containing such groups, only moving to broader screening when failure to establish good candidate leads is encountered. This may be regarded as screening sets of diverse molecules, all of which contain an extra, minimal, conformity element. Having designed the conformity element, further diversity can be built in as above.

A second element of structural conformity may be distance based, with particular pharmacophoric combinations being required to be displayable by the molecule. Such is often the case when lead compounds are available from other sources, and it is required to produce new molecular variants of the active pharmacophore. A screening set of Endothelin-A antagonists was assembled with such a structural conformity [21].

4. COMMERCIAL DIVERSITY

Whatever the diversity methodology chosen, and many are described in the later chapters of this book, there is one further diversity element that is perhaps overwhelming. This is to ensure that compounds synthesised should be novel and patentable. Diversity in this sense (commercial diversity) will give the pharmaceutical house an immense advantage in intellectual property. The real-time marriage of drug design and the patentability of the designed compounds should be addressed.

5. CONCLUSION

The thesis of this chapter is that CADD has a major role to play in the design of the molecules of the future, and that considerations of pharmacological conformity - that the molecules designed have the best chance of being fit-for-purpose - should be placed before considerations of how diverse molecules are from one another.

ACKNOWLEDGEMENTS

The author would like to thank Dr. Stephen Pickett, Dr. David Clark, Dr. Richard Lewis and Dr. Bryan Slater for many discussions and valuable criticisms on the content of this chapter.

REFERENCES:

1. Good, A.C. and Lewis, R.A. New Methodology for Profiling Combinatorial Libraries and Screening Sets: Cleaning Up the Design Process with HARPick. *J. Med. Chem.*, 1997, **40**, 3926-3936.
2. Fecik, R.A., Frank, K.E., Gentry, E.J., Menon, S.R., Mitscher, L.A. and Telikepalli, H. The search for orally active medications through combinatorial chemistry. *Medicinal Research Reviews*, 1998, **18**, 149-185.
3. Messer, M. Traditional or Pragmatic Research. In *Drug Design, Fact or Fantasy*?, Ed. Jolles, G. and Wooldridge, K.R.H., 1984, Academic Press, London.
4. Taylor, J.B. and Kennewell, P.D. *Modern Medicinal Chemistry*, Ellis Horwood, London (1993).

5. Davies, K. Using Pharmacophore Diversity to Select Molecules to Test from Commercial Catalogues. In *Molecular Diversity and Combinatorial Chemistry: Libraries and Drug Discovery*, Ed. Chaiken, I.M. and K.D. Janda, K.D., 1996, American Chemical Society, Washington DC, pp 309-316.
6. Lewis, R.A., Mason, J.S. and McLay, I. Similarity measures for rational set selection and analysis of combinatorial libraries: the Diverse Property-Derived (DPD) approach. *J. Chem. Inf. Comput. Sci.,* 37, 599-614 (1997).
7. Lipinski, C.A., Lombardo, F., Dominy, B.W. and Feeney, P.J. Experimental and computational approaches to estimate solubility and permeability in drug discovery and development settings. *Adv. Drug Delivery Research*, 1997, **23**, 3-25.
8. Pickett, S.D. unpublished data
9. Hirschmann, R., Smith, A.B. and Sprengeler, P.A. Some Interactions of Macromolecules with Low Molecular Weight Ligands. Recent Advances in Peptidomimetic Research. In *New Perspectives in Drug Design*, Eds. Dean, P.M., Jolles, G. and Newton, C.G., 1995, Academic Press, London.
10. Palm, K., Stenburg, P., Luthman, K. and Artursson, P. Polar Molecular Surface Properties Predict the Intestinal Absorption of Drugs in Humans. *Pharma. Res.*, 1997, **14**, 568-571.
11. Clark, D.E. Rapid calculation of polar surface area and its application to the prediction of transport phenomena. 1. Prediction of intestinal absorption. *J. Pharm. Sci.*,1999, **88**, in press.
12. Chadha, H.S., Abrahams, M.H. and Mitchell, R.C. Physicochemical Analysis of the Factors Governing the Distribution of Solutes Between Blood and Brain. *Bioorg. Med. Chem. Lett.*, 1994, **21**, 2511-2516.
13. Young, R.C. Mitchell, R.C., Brown, T.H., Ganellin, C.R., Griffiths, R., Jones, M., Rana, K.K., Saunders, D., Smith, I.R., Sore, N.E. and Wilks, T.J. Development of a new physicochemical model for brain penetration and its application to the design of centrally acting H2 receptor histamine antagonists. *J. Med. Chem.*, 1988, **31**, 656-671.
14. Lombardo, F., Blake, J.F. and Curatolo, W.J. Computation of brain-blood partitioning of organic solutes via free energy calculations. *J. Med. Chem.*, 1996, 39, 4750-4755.
15. Clark, D.E. Rapid calculation of polar surface area and its application to the prediction of transport phenomena. 1. Prediction of blood-brain barrier penetration. *J. Pharm. Sci.*,1999, **88**, in press.
16. Curatolo, W.J. Physical Chemical Properties of Oral Drug Candidates in the Discovery and Exploratory Development Settings. Presented at the International Conference on Accelerated Drug Discovery and Early Development, Zurich, 1997, Technomic AG, Basel.
17. Yalkowski, S.H. and Pinal R. Estimation of the aqueous solubility of complex organic molecules. *Chemosphere*, 1993, **26**, 1239-1261.
18. Huuskonen, J., Salo, M. and Taskinen, J. Aqueous solubility prediction of drugs based on molecular topology and neural network modeling. *J. Chem. Inf. Comput. Sci.,* 1998, **38**, 450-456.
19. Mitchell B.E. and Jurs, P.C. Prediction of Aqueous Solubility of Organic Compounds from Molecular Structure *J. Chem. Inf. Comput. Sci.,* 1998, **38**, 489-496.
20. Mason J.S. and Pickett, S.D. Partition-based selection. *Perspect. Drug Disc. Des.,*1997, **7/8**, 85-114.
21. Porter, B., Lewis, R.A., Lockey, P.M., McCarthy, C., McLay, I.M., Astles, P.C., Roach, A.G., Brown, T.J., Smith, C., Handscombe, C.M., Walsh, R.J.A., Harper, M.F. and Harris N.V. Selective endothelin A receptor ligands. 1. Discovery and structure-activity of 2,4-disubstituted benzoic acid derivatives *Eur. J. Med. Chem.,* 1997, **32**, 409-423.

Chapter 3

Background Theory of Molecular Diversity

Background Theory of Molecular Diversity

Valerie J. Gillet
University of Sheffield, Western Bank, Sheffield S10 2TN, United Kingdom

Key words: Molecular Diversity, Descriptor Analysis, Library Design

Abstract: Recent developments in the technologies of HTS and combinatorial chemistry have thrown down a challenge to computational chemistry, that of maximising the chemical diversity of the compounds made and screened. This paper examines the theory behind molecular diversity analysis and includes a discussion of most of the common diversity indices, and intermolecular similarity and dissimilarity measures. The extent to which the different approaches to diversity analysis have been validated and compared is reviewed. The effects of designing diverse libraries by analysing product and reagent space are presented, and the issues surrounding the comparison of libraries and databases in diversity space are discussed.

1. INTRODUCTION

During the last few years there has been intense interest in the development of computer-based methods for maximising chemical diversity [1-3]. This interest has arisen as a result of the relatively new technologies of combinatorial chemistry and high throughput screening (HTS). Although HTS has greatly increased the rate of testing of compounds, it is very easy to devise combinatorial chemistry experiments that could generate potentially massive numbers of compounds and that far exceed the capacity of current screening programmes. Thus, there is a great need to be selective about the compounds which are actually synthesised. When designing libraries for lead generation, the assumption is made that maximising the range of structural types within a library will result in a broad range of bioactivity types. Hence, much effort has gone into diversity analysis as an important aspect of library design.

This chapter is concerned with some of the background theory for molecular diversity analysis and includes a discussion of diversity indices, intermolecular similarity and dissimilarity measures. The extent to which the different approaches to diversity analysis have been validated and compared is reviewed. Algorithms for the selection of diverse sets of compounds are covered in detail elsewhere in this book and are mentioned only briefly here. However, consideration is given to whether these algorithms should be applied in reactant or product space.

2. DIVERSITY METRICS

The diversity of a library of compounds denotes the degree of heterogeneity, structural range or dissimilarity within the set of compounds. A number of different diversity metrics have been suggested and all are based, either directly or indirectly, on the concept of intermolecular similarity or distance. Determining the (dis)similarity between two molecules requires firstly that the molecules are represented by appropriate structural descriptors and secondly that a quantitative method of determining the degree of resemblance between the two sets of descriptors exists.

Many different structural descriptors have been developed for similarity searching in chemical databases [4] including 2D fragment based descriptors, 3D descriptors, and descriptors that are based on the physical properties of molecules. More recently, attention has focused on diversity studies and many of the descriptors applied in similarity searching are now being applied in diversity studies. Structural descriptors are basically numerical representations of structures that allow pairwise (dis)similarities between structures to be measured through the use of similarity coefficients. Many diversity metrics have been devised that are based on calculating structural (dis)similarities, some of these are described below.

One of the most commonly used structural descriptors in similarity and diversity studies is that of the 2D fragment bitstring where a molecule is represented by a vector of binary values that indicate the presence or absence of structural features, or fragments, within the molecule. Many different similarity measures or coefficients have been developed to quantify the degree of similarity between such vector based descriptors [5-7]. Usually, the values that can be taken by a coefficient lie in the range 0..1, or they can be normalised to be within this range. A similarity coefficient of 1 indicates that the two molecules are identical with respect to the structural descriptors and a value of 0 indicates that the two molecules are maximally different with respect to the descriptors, for example, they have no fragments in common. A similarity coefficient can be converted to its complementary distance or dissimilarity coefficient by subtraction from unity. Hence, a distance of zero indicates identity with respect to the structural descriptors.

The Tanimoto coefficient is the most commonly used coefficient in chemical similarity work following a study of the performance of a range of similarity coefficients by Willett and Winterman [6]. If two molecules have A and B bits set in their fragment bitstrings, with C of these in common, then the Tanimoto coefficient is:

$$\frac{C}{A+B-C} \tag{1}$$

Other similarity coefficients used in similarity studies include the cosine coefficient, and the Hamming and Euclidean distance measures [7]. Similarity coefficients can also be applied to vectors of attributes where the attributes are real numbers, for example, topological indices or physiochemical properties.

A number of diversity indices are based directly on calculating intermolecular dissimilarities, for example, the normalised sum of pairwise dissimilarities using the cosine coefficient [8,9], and the average nearest neighbour distance using the Tanimoto coefficient [11,12]. A number of diversity selection algorithms are also based on these concepts, for example, dissimilarity-based compound selection (DBCS) [10]; clustering techniques where molecules are first grouped or clustered according to their intermolecular similarities and then a representative subset of compounds can be selected by choosing one of more compounds from each cluster [13]; experimental design methods such as D-optimal design [2, 14]; and stochastic methods such as genetic algorithms (GAs) and simulated annealing that attempt to optimise some diversity index such as average nearest neighbour distance [11, 15-17].

Other diversity indices include the HookSpace index [18] which describes the distribution of functional groups in 3D space within a library of compounds; a count of the number of bits that are set in the union of all the fragment bitstrings in a library [2]; the number of distinct substructures that can be generated from all of the molecules in a library [19]; the number of distinct rings that are present in a database [20]; the number of clusters that result at a given clustering level [21]; and in partitioning methods, where some quantifiable combination of properties is used to define partitions in property space, diversity can be measured by the fraction of partitions that achieve a given occupancy of molecules [22-26].

2.1 Structural Descriptors in Diversity Studies

The use of molecular descriptors for similarity and diversity studies is based on the notion that similar molecules generally produce similar biological effects and hence dissimilar molecules will generally produce dissimilar biological effects [27]. Any molecular descriptor used in library design must be able to distinguish between biologically different molecules. Thus, if a structural descriptor is a good indicator of biological activity then good coverage of biological activity space can be achieved by covering as diverse a range of structural types as possible. In addition, the recognition of biologically similar molecules will allow representative compounds to be chosen and hence any redundancy within a library can be minimised.

An additional consideration on the choice of descriptor for diversity analyses is the speed with which the descriptor can be calculated since diversity studies are often applied to the huge numbers of compounds (potentially millions) that characterise combinatorial libraries. Thus, some computationally expensive descriptors such as field-based descriptors [28] or descriptors derived from quantum mechanics [29] are not appropriate for diversity studies.

Biological activity is known to be determined by a complex range of different properties. Receptor binding is clearly important and is determined by physical properties such as hydrophobicity, electrostatic interactions, the ability to form hydrogen bonds between the receptor and a ligand, and 3D shape. Other important factors include bioavailability, toxicity etc. and other physicochemical properties. Hence, to be effective for diversity studies, structural descriptors should be chosen that attempt to model these various properties. The variety of descriptors used in diversity studies have been reviewed recently by Brown [30] and by Agrafiotis [31]. They include 2D and 3D structural descriptors, topological indices and a range of different physicochemical properties.

2.2 Topological Indices and Physicochemical Properties

Topological indices [32] are single valued integers or real numbers that characterise the bonding patterns in molecules. Many different indices have been developed, such as the molecular connectivity indices, and in diversity studies it is usual to use a large range in order to attempt to describe a structure fully. As many of the indices are correlated, normally some data reduction technique, such as principal components analysis or factor analysis, is used to obtain a smaller set of uncorrelated variables. Topological indices are often combined in diversity studies with other global molecular properties such as calculated logPs, molar refractivity, free energy of solvation and molecular weight [2, 11, 12, 33-35].

Martin *et al.* [2] have developed a diversity measure that uses a combination of logP, topological indices, pairwise similarities calculated from Daylight fingerprints [36], see later, using the Tanimoto coefficient, and atom layer properties based on receptor recognition descriptors. Diverse compounds are selected from reactant pools by using principal components analysis and multidimensional scaling on the properties calculated for each molecule to produce a vector that is input to D-optimal design, an experimental design technique. The objective is to produce molecules that are evenly spread in property space. The method was applied to select representative sets of amines and carboxylic acids in the design of peptoid combinatorial libraries.

Mason *et al.* have developed database partitioning methods that are based on global physicochemical properties. The Diverse Property-Derived (DPD) method [25,26] is based on using molecular/physicochemical properties as descriptors. Six descriptors, that are thought to describe important features in drug/receptor interactions, were chosen that measure hydrophobicity, flexibility, shape, hydrogen-bonding properties and aromatic interactions. Each descriptor was then split into two to four partitions to give 576 theoretical combinations or bins. Eighty-six percent of the bins could be filled by compounds from the RPR corporate compound collection and a diverse screening set was chosen by selecting three compounds from each bin.

2.3 2D fragment-based descriptors

2D fragment-based descriptors were originally developed for substructure search systems [37]. These systems are based on a predefined dictionary of fragments and the presence or absence of the fragments in a structure are recorded in a bitstring. Although they were developed for substructure searching, 2D fragment descriptors have been used successfully in similarity studies, and more recently in diversity studies. Examples of these descriptors are the MACCS structural keys [38], which include atom counts, ring types and counts, augmented atoms, and short linear sequences, and which have been used in diversity studies by Brown and Martin [13, 39] and by McGregor and Pallai [40], and the COUSIN [41] structural keys used by Cheng *et al.* [33].

An alternative approach to fragment based bitstrings are hashed fingerprints such as Daylight and UNITY fingerprints. In Daylight fingerprints [36], all the paths of predefined length in a molecule are generated exhaustively and hashed to several bit positions in a bitstring.

Unity 2D fingerprints [42] are also based on paths and additionally denote the presence of specific functional groups, rings or atoms. Fingerprints have been used in a number of diversity studies, for example [13, 15, 39, 43-46].

Several groups have developed structural descriptors that are based on representing atoms by their physicochemical properties rather than by element types. For example, Kearsley *et al.* [47] have identified atom types as belonging to seven binding property classes: cations, anions, neutral hydrogen bond donors and acceptors, atoms which are both donor and acceptor, hydrophobic atoms and all others. They used two structural descriptors called atom-pairs and topological torsions that are based on these atom types in structure activity relationship studies. Their results showed that the new descriptors based on binding classes are complementary to the original descriptors that are based on element types. Martin *et al.* [2] also identified acidic, basic, hydrogen-bond donor, hydrogen-bond acceptor and aromatic groups for use in diversity studies.

Bauknecht *et al.* [48] and Sadowski *et al.* [49] have also developed molecular descriptors that are based on the physicochemical properties of the atoms in a molecule. They calculate several different electronic properties for each atom in a molecule and then use autocorrelation to generate a fixed-length vector that is independent of the size of the molecule. Autocorrelation was first applied to the topology of a molecular structure by Moreau and Broto [50] using the following function:

$$A(d) = \sum_{i,j} P_i P_j \tag{2}$$

where $A(d)$ is the autocorrelation coefficient, P_i and P_j are atomic properties on atoms i and j, respectively, and d is the topological distance between the two atoms measured in bonds along the shortest path. Bauknecht *et al.* calculated their autocorrelation vector using seven atomic properties over seven topological distances to give a vector with 49 dimensions which was then projected into a two-dimensional space via a Kohonen network [51]. The projection resulted in points that are close together in the high-dimensional space occupying the same or adjacent neurons of the network. Using this method, they were able to distinguish between dopamine and benzodiazepine receptor agonists, even when the two sets of agonists were buried within a dataset obtained from a chemical supplier catalogue.

Pearlman [22] has developed novel molecular descriptors called BCUT values for use in diversity studies. They are designed to combine atomic properties with connectivity information in order to define a low-dimensional chemistry space that is relevant to drug-receptor interactions. BCUT values are derived by first creating an association matrix from the connection table for a molecule and then adding atomic properties on the diagonal. The off-diagonals record the connectivity of the molecule. The highest and lowest eigenvalues are then extracted and used as descriptors. For example, a six dimensional space can be defined by using two BCUTS from each of three different matrices: one with atomic-charge related values on the diagonal; a second with atomic polarisabilities on the diagonal and a third with atomic hydrogen-bonding ability on the diagonal. The six dimensional space can then be partitioned and the diversity of a set of compounds determined from their distribution throughout the space. BCUT values have also been developed that encode 3D properties. The same atomic properties are encoded on the diagonals of the matrices and the off-diagonals encode the interatomic distances, calculated from the CONCORD generated 3D structure of the molecule [52].

2.4 3D Descriptors

The fact that receptor binding is a 3D event would suggest that biological activity should be modelled using 3D descriptors, however, there are significant problems associated with the use of such descriptors. These problems arise because, in general, molecules are flexible and they can often adopt a number of different low energy conformations. Also it is often the case that ligands bind to receptors in conformations other than their lowest energy conformations. Thus, some of the issues that have to be considered when calculating 3D descriptors include; the method that is to be used to generate the 3D structures, how conformational flexibility will be handled and which conformers should be included. An additional consideration is the computational effort required, especially when processing large libraries of compounds. Despite these difficulties, 3D descriptors are used in diversity studies.

3D screens were originally designed for 3D substructure searching [37], however, they are now also being used in diversity studies, (cf 2D screens). The screens encode spatial relationships, e.g., distances and angles, between the features in a molecule such as atoms, ring centroids and planes. Distance and angle ranges are specified for each pair of features and each range is then divided into a series of bins by specifying a bin width. For example, a distance range of 0..20Å between two nitrogen atoms might be represented by ten bins each of width 2Å. The 3D features are then represented by a bitstring where the number of bits is equal to the total number of bins for all feature pairs. The presence or absence of feature pairs at certain distance ranges are recorded as for 2D fragments.

Unity 3D rigid screens [42] are based on a single conformation of a molecule, usually that conformation that is generated by CONCORD [52]. Unity 3D flexible screens record all possible distances between the same types of features (atom types, rings and planes) based on the incremental rotation of all the rotatable bonds between the two features. Unity 3D rigid and flexible screens have been used in diversity studies by Brown [13, 39] and by Patterson [43].

Sheridan [53] introduced the concept of pharmacophore keys for 3D database searching. A pharmacophore key is a 3D structural key that is based on the features of a molecule that are thought to have relevance for receptor binding. The features include hydrogen-bond donors, hydrogen-bond acceptors, charged centres, aromatic ring centres and hydrophobic centres. Pharmacophore keys are based on the distances between pairs of these features. Brown and Martin [13] have developed similar keys in-house, where a molecule is reduced to its pharmacophore points and two descriptors are calculated: potential pharmacophore point pairs (PPP Pairs); and potential pharmacophore point triangles (PPP triangles). The PPP Pairs and PPP triangles are then encoded as bitstrings using a hashing scheme similar to that used in the Daylight fingerprints [36].

Similar descriptors have been developed by Pickett *et al.* [23] in the PDQ partitioning method. The PDQ method is a database partitioning method based on the three-point potential pharmacophores present within a molecule as a descriptor. Multiple query 3D structural database searching is performed using a systematic coverage of all pharmacophore types and sizes, using six pharmacophoric points together with six different distance ranges. This gives a total of 5916 valid pharmacophores to be used as queries. The searching takes account of the conformational flexibility of the database molecules. For each compound, information on which pharmacophores can be matched is obtained and for each pharmacophore, the number of times it is matched is stored. The method has been used for database partitioning, pharmacophore identification and library design.

A similar approach based on pharmacophore keys is used in the ChemDiverse software [54]. Here, the key is based on three-point pharmacophores generated for seven features over 32 distances. This gives over 2 million theoretical combinations; however, this number can be reduced by geometric and symmetry considerations. The key marks the presence or absence of the pharmacophores within the collection and because of its size it is normally used to represent a whole library of compounds, although in principle it can also be used to represent a single compound.

Chapman [55] describes a method for selecting a diverse set of compounds that is based on 3D similarity. The diversity of a set of compounds is computed from the similarities between all conformers in the dataset, where multiple conformers are generated for each structure. The similarity between two conformers is determined by aligning them and measuring how well they can be superimposed in terms of steric bulk and polar functionalities. A diverse subset is built by adding one compound at a time and the compound that would contribute the most diversity to the subset is chosen in each step. The high computational cost of this method restricts its use to small datasets.

2.5 Validation of structural descriptors

Despite the many different approaches to diversity analysis, little has yet been done to determine which methods are the best. The studies that have been carried out so far to validate the effectiveness of different structural descriptors in diversity analysis is normally done using simulated property prediction experiments and by examining the coverage of different bioactivity types in the diverse subsets selected. The most extensive studies have been performed by Brown and Martin [13, 39] and by Matter [45].

Brown and Martin [13] compared a range of structural descriptors using different clustering methods and assessed their effectiveness according to how well they were able to distinguish between active and inactive compounds. The effectiveness of the descriptors was found to be, in decreasing order, MACCS and SSKEYS [10] structural keys > Daylight and Unity hashed fingerprints > 3D PPP pairs > Unity 3D rigid and flexible > 3D PPP triangles. The most effective descriptor was the 2D MACCS keys even though it was designed for optimum screenout during substructure search, rather than for similarity searching. However, the poor performance of the 3D descriptors may be due to the fact that only a single conformation was included for each compound.

Brown and Martin [39] also investigated the performance of a number of different descriptors in simulated property prediction experiments. Each descriptor was assessed by its ability to accurately predict the property of a structure from the known values of other structures that were calculated to be similar to it, using the descriptor in question. The predicted properties included measured logP values and calculated properties that explored the shape and flexibility of the molecules, including the numbers of hydrogen-bond donors and acceptors within a molecule. Their results showed the same trend in descriptor effectiveness as their previous study.

Matter [45] has also validated a range of 2D and 3D structural descriptors for their ability to predict biological activity and for their ability to be able to sample structurally and biologically diverse datasets effectively. The descriptors examined included: Unity 2D fingerprints [42], atom-pairs [47], topological 2D descriptors including electrotopological state values [32], molecular connectivity indices [32], molecular shape indices [32] and topological symmetry indices [32], 2D and 3D autocorrelation functions containing electrostatic and lipophilic properties [48,49], flexible 3D fingerprints [56], molecular steric fields based on the comparative molecular field analysis (CoMFA) technique [57] (this descriptor is only suitable for small quantitative structure-activity (QSAR) datasets since it requires that the molecules be superimposed), and WHIM indices (weighted holistic invariant molecular indices) [58] which contain information about the 3D structure of a molecule in terms of size, shape, symmetry and atoms distribution. The 3D autocorrelation functions and WHIM indices do not require molecular superimpositions, and hence are more suitable for large diverse datasets than many of the descriptors that have been described for 3D QSAR. The compound selection techniques used were maximum dissimilarity and clustering. The results showed the 2D fingerprint-based descriptors to be most effective in selecting representative subsets of bioactive compounds, in agreement with the conclusions reached by Brown and Martin.

Patterson *et al.* [43] introduced the concept of neighbourhood behaviour for molecular descriptors where a descriptor that shows neighbourhood behaviour is a good predictor of biological activity. The differences in descriptor values were compared with differences in biological activities for a number of related compounds. Neighbourhood behaviour is identified by plotting the similarity between pairs of compounds against the differences in their activity. A plot that has a characteristic trapezoidal distribution indicates good neighbourhood behaviour. They examined 11 descriptors applied to 20 datasets. Their descriptors included: Unity 2D fingerprints calculated for the whole molecules; Unity 2D fingerprints calculated for side chains only, i.e., if there was a template common to all molecules in the dataset it was removed prior to calculating the fingerprints; 3D CoMFA fields; topological indices; connectivity indices; atom pairs; and autocorrelation indices. Their results showed that 3D CoMFA fields and 2D fingerprints calculated for side chains far out-performed physicochemical properties such as logP and molar refractivity which showed no useful behaviour. However, as mentioned above, use of the 3D CoMFA fields is restricted to small QSAR type datasets, and is not appropriate for large combinatorial libraries, since it requires that the molecules are superimposed. A limitation of this validation study is that it was applied to small QSAR datasets only and the results may not be transferable to the large datasets that are characteristic of combinatorial libraries.

Kauver *et al.* [59] have developed a structural descriptor, called an affinity fingerprint, that is based on the binding affinities of a molecule for a set of reference proteins. The fingerprint is a vector of IC50's for each protein. They compared the affinity fingerprint with a set of 123 physicochemical property values calculated using Molconn-X [32]. They found that the space covered by affinity fingerprints was complementary to that covered by the physicochemical properties and hence concluded that affinity fingerprints are useful descriptors of biological activity.

Rose *et al.* [35] have shown that additive bulk properties such as logP and molar refractivity, 2D structural properties including fingerprints and connectivity indices, and 3D parameters such as dipole moments and moments of inertia, each describe different aspects of the chemical properties of molecules and hence are complementary to one another.

Briem and Kuntz [60] compared similarity searching using Daylight fingerprints [36] with fingerprints generated using the DOCK program. The DOCK fingerprints are based on shape and electrostatic properties. The Daylight 2D descriptors performed better than the DOCK 3D descriptors at identifying known active compounds, thus, providing more evidence in support of the use of 2D descriptors in (dis)similarity studies. The DOCK descriptors were, however, found to be complementary to the 2D descriptors.

3. RANDOM OR RATIONAL?

Given the enormous effort that is being expended on designing diverse libraries, it is of great importance to validate the rational methods for selecting diverse compound subsets compared with selecting compounds at random. The assumption made in these analyses is that rationally designed subsets will contain more diverse sets of compounds, and hence a wider coverage of bioactivity space, than will randomly selected subsets. However, there are a number of studies that suggest that computer-based methods are no better than random at selecting bioactive molecules. Young *et al.* [61] compared random and rational selection using a statistical approach. They concluded that in many cases rational selection of compounds will be no better than random selection, especially for non-focused libraries. Taylor [62] simulated cluster-based and dissimilarity-based selection and concluded that cluster-based selection was only marginally better than random and dissimilarity-based selection was worse than random. Spencer [63] also suggests that cluster-based selection is no better than random. More recently, there have been a number of studies that suggest the converse, that is, that rationally designed subsets are more effective at selecting diverse compounds than are randomly selected subsets.

Gillet *et al.* [15] have compared the diversity of rationally selected subsets of compounds with subsets selected at random. They measure diversity as the sum of pairwise dissimilarities using Daylight fingerprints [36] and the cosine coefficient. They investigated three different published libraries and in each case subsets selected by DBCS and GAs were significantly more diverse than libraries selected at random. They also attempted to find the upper and lower bounds on diversity using a number of methods and concluded that DBCS results in near-optimal libraries using their diversity measure.

Snarey *et al.* [46] have compared the relative effectiveness of different DBCS methods with random selection by measuring the extent to which each method results in a set of compounds that exhibit a wide range of biological activities. The effectiveness of the algorithms was quantified by examining the range of biological activities that result from selecting diverse subsets from the World Drugs Index (WDI). Dissimilarity was measured using both UNITY fingerprints and topological indices and the cosine and the Tanimoto coefficient. Their results suggest that both the maximum dissimilarity algorithm described by Lajiness [10] and the sphere exclusion method of Pearlman [22] are more effective than random at selecting compounds associated with a range of bioactivities.

Hassan *et al.* [11] have developed a method for compound selection that is based on Monte Carlo sampling where diversity is optimised using simulated annealing. The aim of Hassan's method was to compare the performance of different diversity metrics for subset selection. The resulting subsets were visualised by embedding the intermolecular distances, defined by the molecules in multi-dimensional property space, into a three-dimensional space. The descriptors studied were topological indices, information content indices that are based on information-theory equations, electronic descriptors such as charge and dipole moment, hydrophobic descriptors such as calculated logP and molar refractivity, and spatial 3D properties that were calculated for a single conformation of each compounds. Principal components analysis was performed to produce 5 to 10 components that explained 90% of the variance. The diversity functions that were optimised were functions of the intermolecular distances in the property space. They found that the MaxMin function, which maximises the minimum squared distance from each molecule to all other molecules, was found to be effective in producing evenly spread compounds.

Pötter and Matter [64] compared maximum dissimilarity methods and hierarchical clustering with random methods for designing compound subsets. The compound selection methods were applied to a database of 1283 compounds extracted from the IndexChemicus 1993 database that contain 55 biological activity classes. A second database consisted of 334 compounds from 11 different QSAR target series. They compared the distribution of actives in randomly chosen subsets with the rationally designed subsets. They found that maximum dissimilarity methods resulted in more diverse subsets (i.e., subsets that covered more biological classes) than did random selections. Another experiment involved a dataset of 138 inhibitors of the angiotensin-converting enzyme. Designed and randomly chosen subsets were used as training sets for 3D-QSAR studies based on CoMFA. The resulting 3D-QSAR models were subsequently used to predict the biological activities of the remaining compounds not included in the training set. They found that the rationally selected subsets led to more stable QSAR models with higher predictive power compared to randomly chosen compounds.

4. DESIGNING DIVERSE LIBRARIES BY ANALYSING PRODUCT SPACE

The many different approaches to compound selection that have been developed have mostly been applied to the selection of diverse reactants for combinatorial libraries. However, a change of one reactant will cause many of the products in the library to change and there is evidence to suggest that optimising diversity in the reactants does not necessarily result in optimised diversity of the resulting combinatorial library [15]. Reactant- versus product-based selection is shown schematically in figure 1. The upper part of the figure shows reactant-based selection where subsets of reactants are selected from each of the two reactant pools and then the subsets of reactants are enumerated. The lower part of the figure shows product-based selection. Here all the reactants are enumerated to form the large virtual library. Subset selection is then performed on the fully enumerated library.

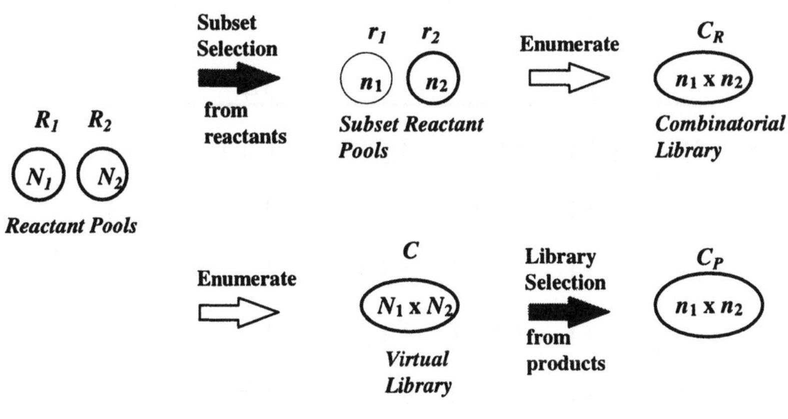

Figure 1. Strategies for library design

Recently several groups have begun to consider selecting molecules in product space. An N component library can be represented by an N dimensional matrix, for example, figure 2 illustrates a 2 dimensional matrix where the rows of the matrix represent the elements in one reactant pool, the columns represent the reactants in the second reactant pool and the elements of the matrix represent the product molecules formed by the combinatorial linking of the reactants in one pool with the reactants in the other.

	$y1$	$y2$	$y3$	$y4$	$y5$
$x1$	$x1\,y1$	$x1\,y2$	$x1\,y3$	$x1\,y4$	$x1\,y5$
$x2$	$x2\,y1$	$x2\,y2$	$x2\,y3$	$x2\,y4$	$x2\,y5$
$x3$	$x3\,y1$	$x3\,y2$	$x3\,y3$	$x3\,y4$	$x3\,y5$
$x4$	$x4\,y1$	$x4\,y2$	$x4\,y3$	$x4\,y4$	$x4\,y5$
$x5$	$x5\,y1$	$x5\,y2$	$x5\,y3$	$x5\,y4$	$x5\,y5$

Figure 2. A matrix showing how reagents are combined combinatorially to give products.

Any of the compound selection methods that have been developed for reactant selection can also be applied to the product library in a process known as *cherry picking*. A subset library selected in this way is shown by the shaded elements of the matrix in figure 3. However, a subset of products selected in this way is very unlikely to be a *combinatorial* library (the compounds in a combinatorial library are the result of combining *all* of the reactants available in one pool with *all* of the reactants in *all* the other pools). Hence, cherry picking is combinatorially inefficient as shown in figure 3 where 7 reactants are required to make the 4 products shown.

	$y1$	$y2$	$y3$	$y4$	$y5$
$x1$	**$x1\,y1$**	$x1\,y2$	$x1\,y3$	$x1\,y4$	$x1\,y5$
$x2$	$x2\,y1$	$x2\,y2$	$x2\,y3$	**$x2\,y4$**	$x2\,y5$
$x3$	$x3\,y1$	$x3\,y2$	$x3\,y3$	$x3\,y4$	$x3\,y5$
$x4$	$x4\,y1$	**$x4\,y2$**	$x4\,y3$	$x4\,y4$	$x4\,y5$
$x5$	$x5\,y1$	$x5\,y2$	$x5\,y3$	**$x5\,y4$**	$x5\,y5$

Figure 3. Selection of a library subset in reagent space, shown as shaded elements.

Several groups have developed genetic algorithms or simulated annealing methods to perform cherry picking in product space. Sheridan and Kearsley [16] have developed a GA for the design of a library of peptoids that are similar to a peptide target. Similarity is measured using atom-pairs as structural descriptor. A chromosome of the GA encodes a single library product that is constructed from fragments extracted from fragment pools. Hence, the GA optimises a population of individual products with no guarantee that the population represents a combinatorial library. In a subsequent step, the fragments that occur frequently in the final products are identified and used to determine a combinatorial synthesis. Zheng *et al.* [65] describe a similar approach that uses topological indices as structural descriptors and simulated annealing as the search algorithm. Liu *et al.* [66] have also developed a similar algorithm, however, in this case a library of product molecules is optimised on diversity rather than on similarity to a target and diversity is measured using steric and electrostatic fields extracted from a CoMFA matrix [57]. A combinatorial library is determined by analysing the frequency of fragments in the final library produced by the GA.

Selecting a combinatorial library from product space is illustrated in figure 4 by the intersection of some of the rows and columns of the matrix that represents the fully enumerated virtual library. Figure 5 illustrates the reordering of the rows and columns of the matrix so that the combinatorial library occupies the top left hand corner of the matrix. Exploring all possible combinatorial libraries is then equivalent to permuting the rows and columns of the matrix in all possible ways. Manipulating a matrix in this way still represents an enormous search space, however, it is possible to search the space efficiently using stochastic optimisation techniques.

	$y1$	$y2$	$y3$	$y4$	$y5$
$x1$	$x1\,y1$	$x1\,y2$	$x1\,y3$	$x1\,y4$	$x1\,y5$
$x2$	$x2\,y1$	$x2\,y2$	$x2\,y3$	$x2\,y4$	$x2\,y5$
$x3$	$x3\,y1$	$x3\,y2$	$x3\,y3$	$x3\,y4$	$x3\,y5$
$x4$	$x4\,y1$	$x4\,y2$	$x4\,y3$	$x4\,y4$	$x4\,y5$
$x5$	$x5\,y1$	$x5\,y2$	$x5\,y3$	$x5\,y4$	$x5\,y5$

Figure 4. Selection of a library subset in product space, shown as shaded elements.

	$y2$	$y5$	$y1$	$y3$	$y4$
$x1$	$x1\,y2$	$x1\,y5$	$x1\,y1$	$x1\,y3$	$x1\,y4$
$x4$	$x4\,y2$	$x4\,y5$	$x4\,y1$	$x4\,y3$	$x4\,y4$
$x2$	$x2\,y2$	$x2\,y5$	$x2\,y1$	$x2\,y3$	$x2\,y4$
$x3$	$x3\,y2$	$x3\,y5$	$x3\,y1$	$x3\,y3$	$x3\,y4$
$x5$	$x5\,y2$	$x5\,y5$	$x5\,y1$	$x5\,y3$	$x5\,y4$

Figure 5. The reordered matrix, showing the selected sublibrary in the top left-hand corner (shaded).

Gillet *et al.* [15] have shown that it is possible to select optimised combinatorial libraries by an analysis of product space. The method is based on a genetic algorithm designed to optimise a multicomponent fitness function. Libraries can be designed to be both diverse and have optimised physical property profiles. Diversity can be measured as the sum of pairwise dissimilarities using the fast O(N) Centroid algorithm [9] which allows selection to be made from very large (~ 1 million compounds) fully enumerated libraries in as little as 2 hours (SG R10000 single processor).

Brown and Martin [67] have recently described a program GALOPED, that is based on a genetic algorithm and that is specifically designed to ease the problems of deconvolution that exist with mixtures. The diversity analysis is based on clustering and requires that the available compounds are preclustered. Thus the selection criteria can be applied to reactant space and also to product space provided that the fully enumerated library is of a manageable size for clustering.

Pickett *et al.* [68] describe a program, DIVSEL, for selecting reactants while taking account of the pharmacophoric diversity that exists in the final products. They describe a 2-component library where the reactants in one pool are fixed and a subset of reactants is to be selected from the second pool. The virtual library is enumerated and a pharmacophore key is generated for each of the product molecules. Reactants are selected from the second pool using a dissimilarity-based compound selection process that represents a candidate reactant by a pharmacophore key that covers an ensemble of products.

Lewis *et al.* [69, 70] have described approaches to product based compound selection that are based on simulated annealing (HARpick) and genetic algorithms and that use pharmacophore keys generated from all possible products. The selection of one product will automatically mean that other products are selected based on the reactants used to generate it. The genetic algorithm has been applied to a benzodiazepine library to select a 4×4×3×2 library from a fully enumerated library that consisted of 1232 structures. The simulated annealing approach was able to give improved results in terms of selecting reactants that give good diversity and that simultaneously satisfy other criteria. These methods are currently limited to relatively small fully enumerated libraries (up to 34 000) compounds.

Analysing the product space of combinatorial libraries can require massive amounts of storage space and it can be very computationally demanding to generate the descriptors for each compound in the virtual library. Downs and Barnard [71] have described an efficient method of generating fingerprint type descriptions of the molecules in a combinatorial library without the need for enumeration of the products from the reactants. This method is based on earlier technology for handling Markush structures.

5. DATABASE COMPARISONS

The ability to compare databases can be extremely useful both for combinatorial library design and also in compound acquisition programs. For example, in library design once a library has been synthesised and tested the screening results can be used to assist in the design of a second library; the second library could be designed to focus in on particular regions of structural space identified as interesting in the first library, or it could be designed to cover a different region of structural space to the first. In compound acquisition, libraries that have minimal overlap with existing in-house collections are generally of greater interest than libraries that do not offer new structural types. Several different methods have been described for comparing databases.

Shemetulskis *et al.* [44] describe a method based on clustering that was used to compare two external databases with a corporate database. Each database was clustered independently using the Jarvis-Patrick method [46]; representative subsets of each database were chosen; and the subsets were then mixed and re-clustered. The number of clusters that contain compounds from only one of the databases was then used as an indication of the degree of overlap between the two databases. A limitation of this approach is the computational effort required to re-cluster the mixed subsets.

Partitioning methods are well suited to database comparisons since the definitions of the partitions are data independent. The PDQ partitioning approach has been used to compare in-house libraries at RPR [24]. The total number of pharmacophores contained within each of three different libraries was calculated and the number of pharmacophores that are common to any two of the three libraries was calculated simply by comparing the cell occupancies of the libraries.

Cummins *et al.* [34] have developed a method for comparing databases that involves characterising the molecules using topological indices and reducing the dimensionality of the descriptor space by factor analysis. Some 60 topological descriptors were selected, using the Molconn-X program, as being as uncorrelated as possible, having a near normal distribution of values and being physically interpretable. Factor analysis was used to reduce the descriptor space to four factors. Several commercially available databases were then compared by partitioning according to the four factors.

Nilakantan *et al.* [20] have developed a method for categorising compounds based on their ring system content. Each ring system in a molecule is hashed to a bitstring according to the atom pairs contained within it. A *ring-cluster* is a seven-letter code that is derived by summing the bitstrings for the ring systems contained in a molecule. Two or more databases can then be compared by comparing the ring-clusters contained in each. The ring-cluster method can also be used to derive a measure of diversity by normalising the number of distinct ring-clusters contained in a database. The normalisation is achieved by dividing by the logarithm of the number of compounds in the database.

Sadowski *et al.* [49] have described the use of 3D autocorrelation vectors that are based on the electrostatic potential measured on the molecular surface of a molecule. The electrostatic potential was measured over 12 different distances giving 12 autocorrelation coefficients per molecule. The vectors were calculated for the molecules in two different combinatorial libraries: a xanthene library and a cubane library. The compounds were then used to train a Kohonen network. The network was successfully able to separate the libraries.

Martin *et al.* [2] describe a method for comparing databases using fragment-based fingerprints. Fingerprints are calculated for each molecule in a database and then ORed to obtain a single fingerprint that represents the whole database. Fingerprints representing different databases can then be compared to give an indication of how similar they are.

Turner *et al.* [72] describe a method for comparing databases that is based on their Centroid algorithm [9]. A centroid is a weighted vector that is derived from fingerprint representations of molecules. The centroid provides an efficient way of calculating the diversity of a database as the sum of pairwise dissimilarities of the molecules contained within it. Combining the centroids from two different databases gives a quantitative measure of the change in diversity that would result from adding the two databases together.

6. CONCLUSIONS

Many different methods have been developed both to measure diversity and to select diverse sets of compounds, however, currently there is no clear picture of which methods are best. To date, some work has been done on comparing the various methods: however, there is a great need for more validation studies to be performed both on the structural descriptors used and on the different compound selection strategies that have been devised. In some cases, the characteristics of the library itself might determine the choice of descriptors and the compound selection methods that can be applied. For example, computationally expensive methods such as 3D pharmacophore methods are limited in the size of libraries that can be handled. Thus for product-based selection, they are currently restricted to handling libraries of tens of thousands of compounds rather than the millions that can be handled using 2D based descriptors.

In diversity analyses, the assumption is made that a structural diverse collection of molecules will lead to a wide coverage of biological activity space, however, it is clear that structurally diverse does not imply "drug-like". There is now increasing interest in the design of libraries that are both diverse and "drug-like" [73-75], for example, through the use of optimisation methods that are based on multi-component fitness functions [76]. Filtering techniques are also important as a way of eliminating undesirable compounds, such as toxic or highly reactive compounds, prior to diversity analyses.

While diversity is of great importance when designing libraries that will be screened across a range of structural targets there is also a growing interest in the design of focused libraries. These are libraries that are designed to span a relatively small region of chemistry space using knowledge that is derived, for example, from a known target structure or from a series of compounds that are known to interact with the target. There is currently much interest in integrating the design of combinatorial libraries with structure-based design techniques [77-79]. This should allow the rational design of combinatorial libraries that are targeted for specific receptors, and should lead to higher hit rates than libraries that are designed using diversity studies alone.

REFERENCES

1. Warr, W. Combinatorial Chemistry and Molecular Diversity, *J. Chem. Inf. Comput. Sci.*, 1997, **37**, 134-140.
2. Martin, E.J., Blaney, J.M., Siani, M.S., Spellmeyer, D.C., Wong, A.K. and Moos, W.H. Measuring Diversity - Experimental Design of Combinatorial Libraries for Drug Discovery. *J. Med. Chem.*, 1995, **38**, 1431-1436.
3. Ellman, J. A. Design, Synthesis, and Evaluation of Small-Molecule Libraries. *Acc. Chem. Res.*, 1996, **29**, 132-143.
4. Downs, G.M. and Willett P. Similarity Searching in Databases of Chemical Structures. In *Reviews in Computational Chemistry*, 1996, **7**, Eds. K.B. Lipkowitz and D.B. Boyd, VCH, New York.
5. Sneath P.H.A. and Sokal R.R. Numerical Taxonomy, 1973, WH Freeman, San Francisco.
6. Willett, P. and Winterman, V. A Comparison of Some Measures for the Determination of Inter-Molecular Structural Similarity. *QSAR*, 1986, **5**, 18-25.
7. Willett, P., Barnard, J.M. and Downs, G.M. Chemical Similarity Searching. *J. Chem. Inf. Comput. Sci.*, 1998, **38**, 983-996.
8. Holliday, J.D. and Willett, P. Definitions of 'Dissimilarity' for Dissimilarity-Based Compound Selection. *J. Biomolecular Screening*, 1996, **1**, 145-151.
9. Holliday, J.D., Ranade S.S. and Willett, P. A Fast Algorithm for Selecting Sets of Dissimilar Structures from Large Chemical Databases. *QSAR*, 1996, **15**, 285-289.
10. Lajiness, M.S. Dissimilarity-Based Compound Selection Techniques. *Perspect. Drug Disc. Des.*, 1997, **7/8**, 65-84.
11. Hassan, M., Bielawski, J.P., Hempel, J.C. and Waldman, M. Optimization and Visualization of Molecular Diversity of Combinatorial Libraries. *Mol. Divers.*, 1996, **2**, 64-74.
12. Hudson, B.D., Hyde, R.M., Rahr, E. and Wood, J. Parameter Based Methods for Compound Selection from Chemical Databases. *QSAR*, 1996, **15**, 285-289.
13. Brown, R.D. and Martin, Y.C. Use of Structure-Activity Data To Compare Structure-Based Clustering Methods and Descriptors for Use in Compound Selection. *J. Chem. Inf. Comput. Sci.*, 1996, **36**, 572-584.
14. Higgs, R.E., Bemis, K.G., Watson, I.A. and Wikel, J.H. Experimental Designs for Selecting Molecules from Large Chemical Databases. *J. Chem. Inf. Comput. Sci.*, 1997, **37**, 861-870.
15. Gillet, V. J., Willett, P. and Bradshaw, J. The Effectiveness of Reactant Pools for Generating Structurally-Diverse Combinatorial Libraries. *J. Chem. Inf. Comput. Sci.*, 1997, **37**, 731-740.
16. Sheridan, R. P. and Kearsley, S. K. Using a Genetic Algorithm to Suggest Combinatorial Libraries. *J. Chem. Inf. Comput. Sci.*, 1995, **35**, 310-320.
17. Agrafiotis, D.K. Stochastic Algorithms for Molecular Diversity. *J. Chem. Inf. Comput. Sci.*, 1997, **37**, 841-851.
18. Boyd, S.M., Beverley, M., Norskov, L. and Hubbard, R.E. Characterising the Geometric Diversity of Functional Groups in Chemical Databases. *J. Comput.-Aided Mol. Des.*, 1995, **9**, 417-424.
19. Bone, R.G.A and Villar, H.O. Exhaustive Enumeration of Molecular Substructures. *J. Comp. Chem.*, 1997, **18**, 86-107.
20. Nilakantan, R., Bauman, N. and Haraki, K.S. Database Diversity Assessment: New Ideas, Concepts, and Tools. *J. Comput.-Aided Mol. Des.*, 1997, **11**, 447-452.

21 Reynolds, C.H., Druker, R. and Pfahler, L.B. Lead Discovery Using Stochastic Cluster Analysis (SCA): A New Method for Clustering Structurally Similar Compounds. *J. Chem. Inf. Comput. Sci.*, 1998, **38**, 305-312.
22. Pearlman, R.S. Novel Software Tools for Addressing Chemical Diversity. *Network Science*, 1996, http://www.netsci.org/Science/Combichem/feature08.html.
23. Pickett, S.D., Mason, J.S. and McLay, I.M. Diversity Profiling and Design Using 3D Pharmacophores: Pharmacophore-Derived Queries (PDQ). *J. Chem. Inf. Comput. Sci.*, 1996, **36**, 1214-1223.
24. Mason, J.S. and Pickett S.D. Partition-Based Selection. *Perspect. Drug Disc. Des.*, 1997, **7/8**, 85-114.
25. Mason, J.S., McLay, I.M. and Lewis, R.A. Applications of Computer-Aided Drug Design Techniques to Lead Generation. In *New Perspectives in Drug Design*, 1994, Eds. P.M Dean, G. Jolles and C.G. Newton, Academic Press, London, pp. 225-253.
26. Lewis, R.A., Mason, J.S. and McLay, I.M. Similarity Measures for Rational Set Selection and Analysis of Combinatorial Libraries: The Diverse Property-Derived (DPD) Approach. *J. Chem. Inf. Comput. Sci.*, 1997, **37**, 599-614.
27. Johnson, M.A. and Maggiora, G.M. Concepts and Applications of Molecular Similarity, 1990, John Wiley, New York.
28. Wild, D.J. and Willett, P. Similarity Searching in Files of Three-Dimensional Chemical Structures: Alignment of Molecular Electrostatic Potentials with a Genetic Algorithm. *J. Chem. Inf. Comput. Sci.*, 1996, **36**, 159-167.
29. Downs, G.M., Willett, P. and Fisanick W. Similarity Searching and Clustering of Chemical-Structure Databases Using Molecular Property Data. *J. Chem. Inf. Comput. Sci.*, 1994, **34**, 1094-1102.
30. Brown, R.D. Descriptors for diversity analysis. *Perspect. Drug Disc. Des.*, 1997, **7**, 31-49.
31. Agrafiotis, D.K. Molecular Diversity. In *Encyclopedia of Computational Chemistry*, Ed. P. v. R. Schleyer, 1998, Wiley.
32. Molconn-X, 1993, available from Hall Associates, Massachusetts.
33. Cheng, C., Maggiora, G., Lajiness, M. and Johnson, M. Four Association Coefficients for Relating Molecular Similarity Measures. *J. Chem. Inf. Comput. Sci.*, 1996, **36**, 909-915.
34. Cummins, D.J., Andrews, C.W., Bentley, J.A. and Cory, M. Molecular diversity in chemical databases: comparison of medicinal chemistry knowledge bases and databases of commercially available compounds. *J. Chem. Inf. Comput. Sci.*, 1996, **36**, 750-763.
35. Rose, V.S., Rahr, E. and Hudson, B.D. The Use of Procrustes Analysis to Compare Different Property Sets for the Characterisation of a Diverse Set of Compounds. *QSAR*, 1994, **13**, 152-158.
36. Daylight Chemical Information Systems, Inc., Mission Viejo, CA, USA.
37. Barnard, J.M. Substructure Searching Methods - Old and New. *J. Chem. Inf. Comput. Sci.*, 1993, **33**, 532-538.
38. MACCS II. Molecular Design Ltd., San Leandro, CA.
39. Brown, R.D. and Martin Y.C. The Information Content of 2D and 3D Structural Descriptors Relevant to Ligand Binding. *J. Chem. Inf. Comput. Sci.*, 1997, **37**, 1-9.
40. McGregor, M.J. and Pallai, P.V. Clustering of Large Databases of Compounds: Using the MDL "Keys" as Structural Descriptors. *J. Chem. Inf. Comput. Sci.*, 1997, **37**, 443-448.
41 Hagadone, T.R. Molecular Substructure Similarity Searching - Efficient Retrieval in 2-Dimensional Structure Databases. *J. Chem. Inf. Comput. Sci.*, 1992, **32**, 515-521.
42. UNITY Chemical Information Software. Tripos Inc., 1699 Hanley Rd., St. Louis, MO 63144.

43. Patterson, D.E., Cramer, R.D., Ferguson, A.M., Clark, R.D. and Weingerger, L.E. Neighbourhood Behaviour: A Useful Concept for Validation of "Molecular Diversity" Descriptors. *J. Med. Chem.,* 1996, **39**, 3049-3059.
44. Shemetulskis, N.E., Dunbar, J.B., Dunbar B.W., Moreland D.W. and Humblet, C. Enhancing the Diversity of a Corporate Database Using Chemical Database Clustering and Analysis. *J. Comput.-Aided Mol. Des.*, 1995, **9**, 407-416.
45. Matter H. Selecting Optimally Diverse Compounds from Structural Databases: A Validation Study of Two-Dimensional and Three-Dimensional Molecular Descriptors. *J. Med. Chem.*, 1997, **40**, 1219-1229.
46. Snarey, M., Terrett, N.K., Willett, P. and Wilton D.J. Comparison of Algorithms for Dissimilarity-Based Compound Selection. *J. Mol. Graph. Modelling*, 1997, **15**, 372-385.
47. Kearsley, S.K., Sallamack, S., Fluder, E.M., Andose, J.D., Mosley, R.T. and Sheridan, R.P. Chemical Similarity Using Physiochemical Property Descriptors. *J. Chem. Inf. Comput. Sci.*, 1996, **36**, 118-127.
48. Bauknecht, H., Zell, A., Bayer, H., Levi, P., Wagener, M., Sadowski, J. and Gasteiger, J. Locating Biologically Active Compounds in Medium-Sized Heterogeneous Datasets by Topological Autocorrelation Vectors: Dopamine and Benzodiazepine Agonists. *J. Chem. Inf. Comput. Sci.*, 1996, **36**, 1205-1213.
49. Sadowski, J., Wagener, M. and Gasteiger, J. (1995) Assessing Similarity and Diversity of Combinatorial Libraries by Spatial Autocorrelation Functions and Neural Networks. *Angew. Chem. Int. Ed. Engl.*, 1995, **34**, 2674-2677.
50. Moreau, G. and Broto, P. Autocorrelation of Molecular Structures: Application to SAR studies. *Nouv. J. Chim.*, 1980, **4**, 757-764.
51. Kohonen, T. Self-Organization and Associative memory, 3rd Ed, 1989, Springer, Berlin.
52. CONCORD. A Program for the Rapid Generation of High Quality Approximate 3-Dimensional Molecular Structures. The University of Texas at Austin and Tripos Inc, St. Louis, MO 63144.
53. Sheridan, R.P., Nilikantan, R., Rusinko, A., Bauman, N., Haraki, K. and Venkataraghavan, R. 3DSEARCH: A System for Three-Dimensional Substructure Searching. *J. Chem. Inf. Comput. Sci.*, 1989, **29**, 255-260.
54. ChemDiverse. Oxford Molecular Group, Oxford Science Park, Oxford, UK.
55. Chapman, D.J. The Measurement of Molecular Diversity: A Three-Dimensional Approach. *J. Comput-Aided Mol. Des.*, 1996, **10**, 501-512.
56. SYBYL Molecular Modelling Package. Tripos Inc., St. Louis, MO 63144.
57. Cramer, R.D., Patterson, D.E. and Bunce, J.D. Comparative Molecular Field Analysis (CoMFA). 1. Effect of Shape on Binding of Steroids to Carrier Proteins. *J. Am. Chem. Soc.*, 1988, **110**, 5959-5967.
58. Todeschini, R.; Lasagni, M.; Marengo, E. New Molecular Descriptors for 2D and 3D Structures - Theory. *J. Chemom.*, 1994, **8**, 263-272.
59. Kauver, L.M., Higgins, D.L., Villar, H.O., Sportsman, J.R., Engqvist-Goldstein, A., Bukar, R., Bauer, K.E., Dilley, H. and Rocke, D.M. Predicting Ligand-Binding to Proteins by Affinity Fingerprinting. *Chem. Biol.*, 1995, **2**, 107-118.
60. Briem, H. and Kuntz, I.D. Molecular Similarity Based on Dock-Generated Fingerprints. *J. Med. Chem.*, 1996, **39**, 3401-3408.
61. Young, S.S., Farmen, M. and Rusinko, A. Random Versus Rational. Which is Better for General Compound Screening? *Network Sci.* (Electronic publication) 1996, **2**.
62. Taylor, R. Simulation analysis of experimental design strategies for screening random compounds as potential new drugs and agrochemcials. *J. Chem. Inf. Comput. Sci.*, 1995, **35**, 59-67.

63. Spencer, R.W. Diversity Analysis in High Throughput Screening. *J. Biomolecular Screening*, 1997, **2**, 69-70.
64. Pötter, T. Matter, H. Random or Rational Design? Evaluation of Diverse Compound Subsets from Chemical Structure Databases. *J. Med. Chem.*, 1998, **41**, 478-488.
65. Zheng, W., Cho, S.J. and Tropsha, A. Rational Combinatorial Library Design. 1. Focus-2D: A New Approach to the Design of Targetted Combinatorial Chemical Libraries. *J. Chem. Inf. Comput. Sci.*, 1998, **38**, 251-258.
66. Liu, D., Jiang, H., Chen, K. and Ji, R. A New Approach to Design Virtual Combinatorial Library with Genetic Algorithm Based on 3D Grid Property. *J. Chem. Inf. Comput. Sci.*, 1998, **38**, 233-242.
67. Brown, R.D. and Martin, Y.C. Designing Combinatorial Library Mixtures Using a Genetic Algorithm. *J. Med. Chem.*, 1997, **40**, 2304-2313.
68. Pickett, S. D., Luttmann, C., Guerin, V., Laoui, A. and James, E. DIVSEL and COMPLIB - Strategies for the Design and Comparison on Combinatorial Libraries using Pharmacophoric Descriptors. *J. Chem. Inf. Comput. Sci.*, 1998, **38**, 144-150.
69. Lewis, R.A., Good, A.C. and Pickett, S.D. Quantification of Molecular Similarity and Its Application to Combinatorial Chemistry. In *Computer-Assisted Lead Finding and Optimization: Current Tools for Medicinal Chemistry* Eds. van de Waterbeemd, H., Testa, B. and Folkers, G., Wiley-VCH:Weinheim, 1997, pp. 135-156.
70. Good, A.C. and Lewis, R.A. New Methodology for Profiling Combinatorial Libraries and Screening Sets: Cleaning Up the Design Process with HARPick. *J.Med.Chem.*, 1997, **40**, 3926-3936.
71. Downs, G.M. and Barnard, J.M. Techniques for Generating Descriptive Fingerprints in Combinatorial Libraries. *J. Chem. Inf. Comput. Sci.*, 1997, **37**, 59-61.
72. Turner, D.B., Tyrrell, S.M. and Willett, P. Rapid Quantification of Molecular Diversity for Selective Database Acquisition. *J. Chem. Inf. Comput. Sci.*, 1997, **37**, 18-22.
73. Gillet, V.J., Willett, P. and Bradshaw, J. Identification of Biological Activity Profiles Using Substructural Analysis And Genetic Algorithms. *J. Chem. Inf. Comput. Sci.*, 1998, **38**, 165-179.
74. Ajay, Walters, W.P. and Murcko, M. Can We Learn to Distinguish between "Drug-like" and "Nondrug-like" Molecules? *J. Med. Chem.*, 1998, **41**, 3314-3324.
75. Sadowski, J. and Kubinyi, H. A Scoring Scheme for Discriminating between Drugs and Nondrugs. *J. Med. Chem.*, 1998, **41**, 3325-3329.
76. Gillet, V.J., Willett P., Bradshaw, J. and Green D.V.S. Selecting combinatorial libraries to optimise diversity and physical properties. *J. Chem. Inf. Comput. Sci.*, 1999, **39**, 169-177.
77. Jones, G., Willett, P., Glen, R.C., Leach, A.R. and Taylor, R. Further Development of a Genetic Algorithm for Ligand Docking and its Application to Screening Combinatorial Libraries. *ACS Symposium Series*, in press.
78. Murray, C.W., Clark, D.E., Auton, T.R., Firth, M.A., Li, J., Sykes, R.A., Waszkowycz, B., Westhead, D.R. and Young, S.C. PRO_SELECT: Combining Structure-Based Drug Design and Combinatorial Chemistry for Rapid Lead Discovery. 1. Technology. *J. Comput.-Aided Mol. Des.*, 1997, **11**, 193-207.
79. Graybill, T.L., Agrafiotis, D.K., Bone, R., Illig, C.R., Jaeger, E.P., Locke, K.T., Lu, T., Salvino, J.M., Soll, R.M., Spurlino, J.C., Subasinghe, N., Tomczuk, B.E. and Salemme, F.R. Enhancing the Drug Discovery Process by Integration of High-Throughput Chemistry and Structure-Based Drug Design. In *Molecular Diversity and Combinatorial Chemistry*, Eds. I.E. Chaiken and K.D. Janda, 1996, ACS, Washington D.C.

Chapter 4

Absolute vs Relative Similarity and Diversity

The Partitioning Approach to relative and absolute diversity

Jonathan S. Mason
Bristol-Myers Squibb, PO Box 4000, Princeton NJ08543 USA

Key words: diversity, similarity, pharmacophores, chemistry space, library design

Abstract: Similarity and diversity methods play an important role in new applications such as virtual screening, combinatorial library design and the analysis of hits from high throughput screening. This paper describes an approach that measures 'relative' similarity and diversity between chemical objects, in contrast to the use of the concept of a total or 'absolute' reference space. The approach is elucidated using the multiple potential 3D pharmacophores method (a modification to the Chem-X/ChemDiverse method), which can be used for both ligands and protein sites. The use of 'receptor-relevant' BCUT chemistry spaces from DiverseSolutions is also discussed.

1. INTRODUCTION

Molecular similarity and diversity methods play an important role in new applications such as virtual screening, combinatorial library design and the analysis of hits from high throughput screening. Similarity and diversity methods can be applied to whole structures or just fragments; in the case of library design, the fragments can be the various reagents and 'scaffolds' (core structures). Some methods, described elsewhere in this book, can be used to determine similarity or diversity in an 'absolute' manner, comparing molecules in terms of the properties of the complete structure. This paper describes an approach that measures 'relative' similarity and diversity, where only a subset of each molecule is considered.

1.1 Multiple potential pharmacophore method

The approach is best illustrated using the multiple potential 3D pharmacophores method [1-7] which is based on functionality available in the Chem-X/ChemDiverse software [8]. A molecule is analysed to determine all potential pharmacophore points it contains, e.g. hydrogen-bond acceptors; the definition of pharmacophore points is explored in more detail in section 2.1. A potential pharmacophore is then defined by a combination of pharmacophore centres and the distances between the centres. For practical reasons, the distances are categorised into bins, to create a finite number of potential pharmacophores. The number of potential pharmacophores is dependent on the number of centre types, the number of distance bins, and the number of points in the pharmacophore. Normally all 3- or 4-point potential pharmacophores for a molecule are calculated during extensive conformational sampling. Because the number of potential pharmacophores is fixed for a particular experiment, the resultant ensemble of potential pharmacophores for a molecule can be encoded as a binary key, depending on whether a potential pharmacophore is expressed or not. The key is thus the sum of all the pharmacophoric shapes for all the acceptable sampled conformations [9]. Keys from several molecules can be combined logically to give new keys that describe similarity or diversity. The multiple potential 3-D pharmacophore method is summarised in Figure 1, which shows some of the numbers involved. Centre-type number 3 can be reserved to describe 'special' features, or used for features such as those that are both donors and acceptors or are tautomeric.

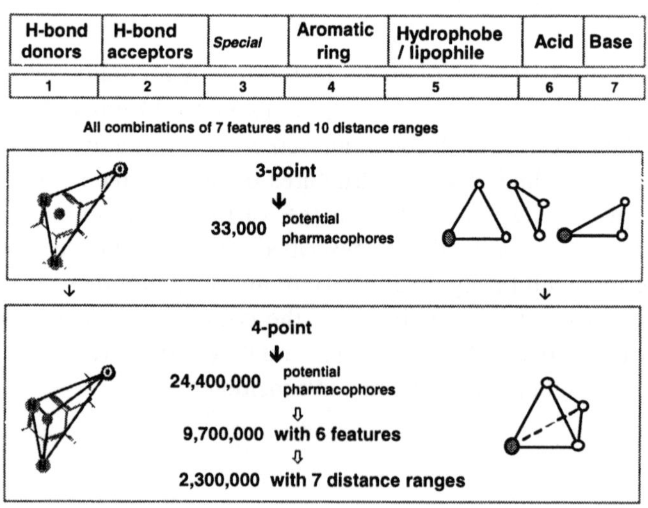

Figure 1. The multiple potential 3-D pharmacophore method.

1.1.1 Relative similarity and diversity

When used for 'relative' similarity and diversity, only potential pharmacophores that contain the defined special centre-type are used. The frame of reference for similarity/diversity studies is thus changed to one that is focused on the feature of interest; distances are now measured relative to this special centre. For example, the special centre could be the centroid of a substructure [10] such as biphenyl tetrazole or diphenylmethane, enabling the calculation and comparison of all 3D pharmacophoric shapes that contain this substructure; the substructure is said to be 'privileged'. For structure-based design, the potential pharmacophores in a site can be restricted to those that contain a specific site point (e.g. in a pocket, or at the entrance to a pocket). In the context of combinatorial library design, the 'relative' measure can be those pharmacophoric shapes that contain a special site-point that represents where the attachment point for a reagent would be. In figure 1, the special point would be centre-type number 3, which can be reserved for this purpose.

Powerful new methods for the design of libraries or molecular similarity, that consider all the pharmacophoric shapes relative to a special point and can use information from known ligands and protein binding sites [11-14], are now possible. As the multiple potential pharmacophore method is a partitioning method, 'missing' diversity can be identified from unfilled partitions, and can be readily visualised as the actual 3- (triangle) or 4- (tetrahedron) point pharmacophores. These are defined in terms of pharmacophore feature types for each point and 3D distances between them, considering all combinations of distances and pharmacophore types (see 1.1), with the restriction for the 'relative' measure that one of the pharmacophore points is the 'special' one. If the diversity is missing in one set of molecules (e.g. a company screening set) but present in another (e.g. a set of known drugs), the molecules that occupy the missing pharmacophore partitions can be visualised (with the relevant 3- or 4-point pharmacophore(s) highlighted).

1.2 DiverseSolutions chemistry space method

The use of 'receptor-relevant' BCUT chemistry spaces from DiverseSolutions (DVS) [15-19] is discussed in section 3.1. This involves the use of a subset of descriptors (atomic/molecular properties) determined to be relevant to discriminate the diversity of a large set of molecules. This method, reported by Pearlman and Smith [19], can be considered as a type of 'relative' similarity and diversity, where the subset of properties that are considered to be important for biological activity are separated from the others.

1.2.1 Relative similarity and diversity

The use of 'relative' similarity and diversity thus enables a focus on the subset of properties considered to be important for binding (biological activity). With the DVS BCUT chemistry space method, the definition of a subset of 'receptor' relevant properties enables similarity-based calculations and designs to not be constrained or diluted with irrelevant information. This allows new molecules to be designed that preserve the desired activity-governing properties, while exploring other aspects of lead optimisation (e.g. bioavailability). With the multiple potential 3D pharmacophore method, this means that pharmacophore coverage relative to special features is not mixed with other pharmacophores that come only from other features in the molecule or protein site. This enables more refined ligand-ligand and ligand-receptor similarity studies to be made (see sections 4.2 and 4.3), and the diversity relative to a feature or substructure of interest to be better explored (see section 5). A combination of these methods can also be used.

2. MULTIPLE POTENTIAL 3D PHARMACOPHORES

A 3- and 4-point multiple potential 3D pharmacophore method for molecular similarity and diversity that rapidly calculates all potential pharmacophores (pharmacophoric shapes) for a molecule, with conformational sampling, or a protein site (using complementary points) is available as an extension to the ChemDiverse/Chem-X software [8]. The method is summarised in figure 1. Customisations of this method to create a 'relative' measure of pharmacophore diversity have been reported [7, 11-14], with applications for the design of combinatorial libraries containing privileged substructures and for ligand-enzyme site selectivity studies.

2.1 Calculation of potential pharmacophores

All 3- and 4-point potential pharmacophores can be calculated, using six pharmacophoric features for each point: hydrogen-bond donor; hydrogen-bond acceptor; aromatic ring centroid; hydrophobic/lipophilic region; acidic centre; basic centre. This can be done for a single conformation, or for an ensemble of accessible conformations; conformational sampling is normally done at search time. A maximum of 32 distance ranges for each of the 3 distances (3-point) and 15 distance ranges for the 6 distances (4-point) are considered, leading to about a half million (3-point) to 100 million (4-point) pharmacophoric shapes being considered for each molecule. This information is stored in a pharmacophore 'key'. All accessible geometric combinations are calculated, but effects such as the triangle inequality rule greatly reduces the actual number of potential pharmacophores from the theoretical combination of all distance ranges (e.g. for 3-point pharmacophores and 32 distances: >2 million reduces to ~850,000 for combinations from seven possible features).

This produces a pharmacophore key or 'signature' that provides a powerful measure for 'absolute' diversity or similarity, calculable for both a ligand and a protein site, that has a consistent frame of reference for comparing any number of molecules and for comparing molecules to protein sites. Pharmacophore keys can be compared using logical operations (OR, AND, NOT) within Chem-X, and can be outputted in an ASCII or binary format, and analysed by other programs.

2.1.1 Calculation for ligands

The multiple potential pharmacophores are calculated for ligands by automatically assigning one of the six or seven pharmacophoric features to relevant atoms (identified by their 'atom-type') or to 'dummy' atoms that represent a set of atoms, for example a centroid of an aromatic ring or of a group of atoms that create a hydrophobic region. Atom types are assigned and dummy atoms added by a parameterisation file and fragment database; this is fully user-customisable, enabling the assignment of groups to be readily changed (e.g. whether a nitro group is assigned as a hydrogen-bond acceptor, or to recognise a tetrazole as an acidic group) [3, 20]. Studies have been reported that identify a set of potential pharmacophores that were characteristic for a particular type of molecular target, and highlight the added value of 4-point pharmacophores [3, 7]

2.1.2 Calculation for targets

The multiple potential pharmacophores for targets, such as protein active sites, are defined using complementary site points to exposed features accessible in the site. These site points (see section 2.3) create a hypothetical 'molecule' that interacts with all pharmacophoric regions of the site. Figure 2 illustrates site points that were used for the thrombin site in selectivity studies for three serine protease inhibitors (see section 4.3). The potential pharmacophores are calculated for this 'molecule' just as for any other, except that no conformational sampling is performed; conformational flexibility in the site groups can be incorporated by generating several different complementary site 'molecules', using different conformations of the site, and combining the resultant keys. As the potential pharmacophores are calculated in the same way as for a normal molecule, pharmacophore keys for both ligands and targets can be directly compared, for similarity (ligand-target) or diversity (e.g. to find ligands that explore a site) studies.

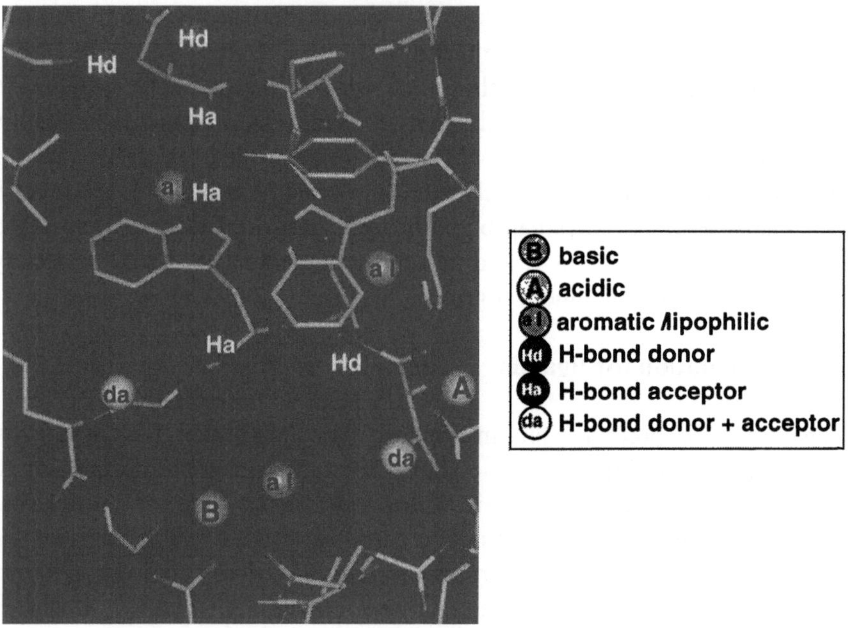

Figure 2. Example site points defined for the thrombin active site.

2.1.3 Definition of features – atom types

Features that are likely to be important for drug-receptor interactions are automatically identified for each molecule by assigning specific atom types (that are associated with one of more features), adding centroid dummy atoms where necessary to represent groups of atoms (for aromatic rings and hydrophobic regions). The six principal features used are hydrogen bond donors, hydrogen bond acceptors, acidic centres (negatively charged at pH 7 pH 7), basic centres (positively charged at pH 7), aromatic ring centroids and hydrophobic regions. Up to seven features are supported in the ChemDiverse software (which refers to them as 'centres'), and this seventh feature can be used to define a 'special' point for 'relative' similarity/diversity studies. By default, in ChemDiverse this extra feature is used for quarternary nitrogens (these can be grouped with the basic feature), and in previously reported studies [1-3], was also used for features that have both hydrogen bond donor and acceptor characteristics (e.g. an OH group, tautomeric atoms).

In a first step, atom types are assigned automatically when reading a new molecule into Chem-X through a parameterisation file and fragment database. Atom types are assigned to distinguish different environments; this process is fully user customisable. Each atom type that could exhibit one of the pharmacophoric features is additionally assigned a feature number. Dummy centroid atoms, which have atom types and thence feature assignments, are added to represent aromatic ring centroids and hydrophobic regions (by an automatic method within Chem-X that uses bond polarities [1]). Molecules are then saved into a database, with assigned atom types. A molecule can be reparameterised before pharmacophoric analysis, using modified assignments.

The need to distinguish atomic environments and to equivalence those with similar properties has been found to be key for molecular similarity/diversity evaluations [1, 7, 20]. Acids and bases are important features, and need special attention; for example, a carboxylic acid, a tetrazole and an acyl sulfonamide have very different 2D structures, but are all acids ionizable at physiological pH and should all have the acidic feature assigned. Tautomeric atoms (e.g. imidazole) should be assigned as donor or acceptor, and deactivated atoms such as di-substituted nitrogens in amides and substituted nitrogens in aromatic systems should not be assigned any features. The method used to assign the atom types is based on an order-dependent database of 'fragments'; either a total connectivity or a bond-order-dependent substructural fragment match is used to assign atom types [1, 20].

2.1.4 Distance ranges

The ChemDiverse method uses predefined ranges for measuring the distances between the points (pharmacophoric features); distances are calculated exactly but stored using this 'binning' scheme, with each distance represented by the bin into whose range it falls. Each pharmacophore for the 4-point method needs six distances to be characterised (to form a tetrahedron), whereas three distances are needed for each 3-point pharmacophore (triangle). All the combinations of features and distances, combined for all the evaluated conformers, are stored in a pharmacophore 'key'.

The default set of distance ranges used lie between 2 and 15 Å with a 1Å interval split for the 4-point method and a 0.1 to 1Å interval split for the 3-

point method (30 ranges). Additional ranges for distances less than or greater than the defined limits are also defined. Thus, 32 (3-point) and 15 (4-point) ranges are used by default. Based on experience with 3D database searching, and taking into account the torsional increments used in the conformational sampling, customised distance ranges were defined and used for both 3- and 4-point calculations in reported work. Longer distances were included and the size of each range was varied so that there is a fixed percentage variance from the mid-point; larger distances thus have larger ranges. 7 or 10 ranges were normally used for the 4-point method, which gives keys of a more manageable size, with a reduced risk of failing to set a bit because of conformational sampling limitations. It appears that adequate resolution/differentiation is obtained with the relatively low number of 7 or 10 distance ranges when 4-point pharmacophores are used [3]. Figure 3 shows the number of potential pharmacophores identified for an endothelin antagonist [21], and illustrates the much larger number of potential pharmacophores that are generated with the 4-point definition. Similarity studies indicate that this extra information is meaningful.

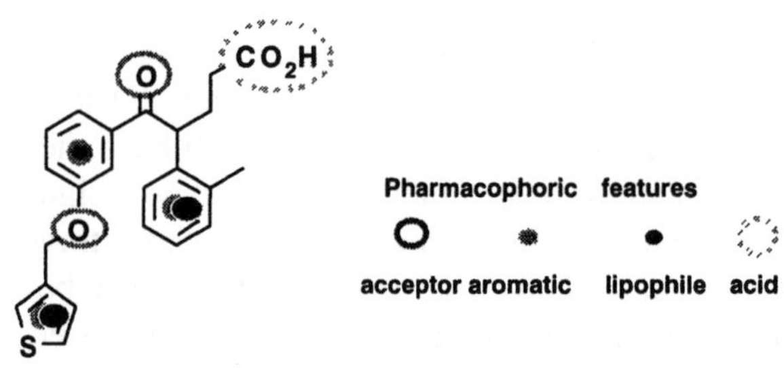

Number of distance ranges	3-point potential pharmacophores	4-point potential pharmacophores
7	422	3298
10	665	6077
16	1665	16200

Figure 3. 3- and 4-point pharmacophores for an endothelin antagonist (6 features/point).

2.1.5 Conformational sampling

The relatively high conformational flexibility of many drug molecules requires that an effective conformational sampling is performed for a pharmacophore-based analysis. The method used in ChemDiverse is based on an explicit "on-the-fly" generation of conformers done at search time. A conformation is accepted or rejected based on a fast evaluation of steric contacts or by using rules. A much slower full energy calculation could be used, but the relevance of such vacuum energies to energies of bound ligands is not clear. The composite key for all accepted conformations of all accessible combinations of 3 or 4 features and geometries (using distance ranges) is stored.

Conformational sampling can be extensively customised, replacing the default systematic rule-based analysis with one that uses systematic or random generation of conformations, with a number of rotamers per bond, and a steric bump check to define acceptability. The default sampling in ChemX is 3 rotamers per single (sp3-sp3) bond, 6 rotamers per alpha (sp2-sp3) bond and 2 per conjugated (sp2-sp2) bond. Reported work has used customised sampling, with 4 rotamers per alpha (sp2-sp3) bond, and 1-4 per conjugated bond (e.g. 1 for mono-substituted amides, 2 for disubstituted amides and 4 for some conjugated amide-aromatic systems). Typical sampling times were reported to be up to 45 sec (on a SGI R4400 chip) or 15 sec (on an SGI R10000 chip) for systematic analysis, and a third of these values for random sampling. As only torsional sampling is performed, it is important to use high quality structures with standard bond angles, such as from the program CONCORD [22]; relaxing to a particular conformer can cause falsely high energy structures for rotamers where the relaxation works in the wrong sense.

2.1.6 Chirality

An advantage of 4-point pharmacophores over 3-point ones is the ability to distinguish chirality, and a flag for this can be optionally added to the 4-point pharmacophore calculation; distances alone cannot distinguish chirality. With this option, which increases the size of the keys, separate bits in the pharmacophore key are set for the two 'enantiomers' for all chiral combinations of features. When chiral information is available, for example when using complementary pharmacophores from a protein active site, this is an important option, and from a theoretical perspective, it clearly is a requisite for an effective similarity/diversity measure given the large difference in biological activity that can be observed between drug enantiomers.

2.1.7 Frequency Count

Another option is the ability to add a count per pharmacophore. The count can be set to be either per molecule, whereby a maximum of one is added to the count irrespective of the number of matching conformations, or per matching conformation. Additional pharmacophore key bits are set to count the number of occurrences and the logical operators available to compare keys are modified accordingly; the key size is increased significantly with a count, depending on the maximum stored count defined. This option has been reported to be particularly useful with pharmacophore keys calculated for sets of molecules sharing a common type of biological activity. Results reported with 7TM-GPCR ligands showed that by using even a very simple count, it was possible to delineate a set of potential pharmacophores that appear to be enriched in this class of compounds [3, 7].

2.1.8 Quality checks

ChemDiverse supports two optional fast "quality" checks that can be applied to potential pharmacophores before they are added to the key. These checks exclude pharmacophores that are either too small relative to the whole molecule ('volume' check) or are potentially inaccessible to a receptor interaction ('accessibility' check).

Based on an empirical formula, the 'volume' check compares the area (3-point) or volume (4-point) of the potential pharmacophore with the heavy atom count for the molecule. This can exclude pharmacophores that have relatively small size when compared to the molecule size, for example a pharmacophore only involving a single residue in a tetrapeptide molecule. Although the estimate is very approximate, using only the heavy atom count, it appears to be a useful filter.

The 'accessibility' check eliminates pharmacophores where the putative interacting hydrogen atom or lone pair points within the triangle of the 3-point pharmacophore. The validation of such a filter is less clear, and it has not been used in the reported studies.

2.2 Calculation of relative potential pharmacophores

A powerful extension to the potential pharmacophore method has been developed, in which one of the points is forced to contain a 'special' pharmacophore feature, as illustrated in figure 4. All the potential pharmacophores in the pharmacophore key must contain this feature, thus making it possible to reference the pharmacophoric shapes of the molecule relative to the 'special' feature. This gives an internally referenced or 'relative' measure of molecular similarity/diversity. The 'special' feature can be assigned to any atom-type or site-point, or to special dummy atoms, such as those added as centroids of 'privileged' substructures [7, 10]. With one of the points being reserved for this 'special' feature, it would seem even more necessary to use the 4-point definition to capture enough of the pharmacophoric and shape properties. A customised geometry type file is required to store only the pharmacophores containing the 'special' feature.

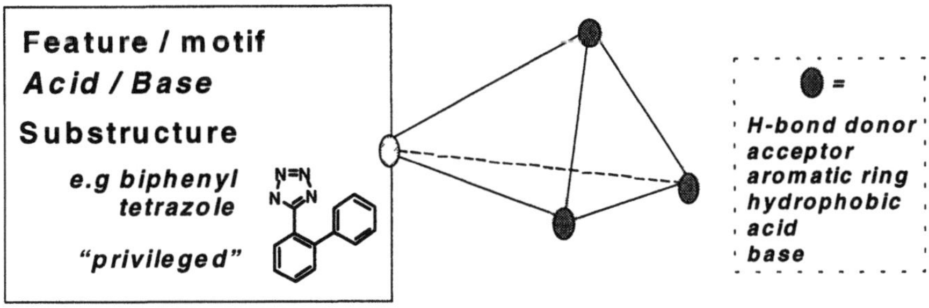

Figure 4. 'Relative' pharmacophore definition.

The 'special' feature uses the extra feature possible in ChemDiverse in addition to the six discussed in section 2.1. The number associated with this extra feature, not otherwise assigned to any other atom type, is simply assigned to the atom type of the atom that is desired to be coded as 'special'. For a group or substructure of interest, a special dummy atom can be added, with a unique atom type that is assigned the 'special' feature number. This can be achieved by adding a fragment to the parameterisation database [1, 20]. An existing atom type can be used, as long as it is only assigned to atoms in the 'special' feature. Multiple assignments may however be acceptable, for example to measure diversity relative to an acidic feature; in this case, more than one acidic function could be present in a molecule. Figure 5 illustrates a sample 'privileged' potential pharmacophore for a biphenyltetrazole-containing compound.

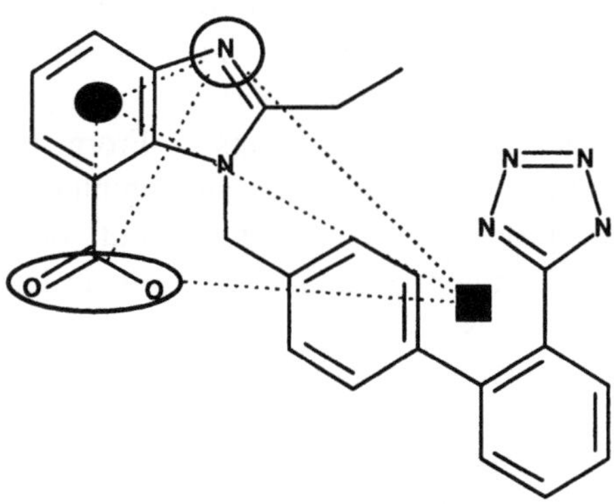

Figure 5. Example of a 'relative' pharmacophore, with the 'privileged' biphenyltetrazole substructure as the 'special' point (shown as a square), and the connections to the other centres (dotted lines).

For site-points, one of the points just needs to be manually reassigned to an otherwise unused atom type, to which the 'special' feature is assigned. Figure 6 illustrates 'relative' pharmacophores in a site context.

Figure 6. An example of a relative pharmacophore given in the context of structure(site)-based design.

2.3 Generation of complementary pharmacophores for protein sites

Potential 3 or 4-point pharmacophores for a target site, such as an enzyme active site or a receptor site can be calculated using complementary site points. The site points can be generated by several different methods; the method in Chem-X/ChemProtein uses a template file where complementary site points for all residue types are defined and positioned relative to the residue atoms. A hydrogen bond donor atom would thus be placed complementary to an accessible backbone carbonyl oxygen atom whilst a hydrogen bond acceptor would be placed to face an amide protein backbone nitrogen. The ChemProtein/Receptor Screening method in Chem-X uses by default a simple geometric addition of "dummy" atoms with associated feature ('centre') types, based on a customizable database of amino acid fragments, for which site points have been positioned.

The GRID [23] program performs energetic surveys of the site, using a wide variety of probe atoms or groups. From the resultant maps, energetically contoured, it is possible to locate complementary site points for hydrogen bond acceptor, donor, acceptor and donor, acidic, basic, hydrophobic or aromatic interactions. An automatic location of "dummy" atoms at energetic minima is also available as a GRID program option. The positions of potential interacting atoms or groups could be verified and optimized using a force-field molecular dynamics simulation and minimization. A method based on crystallographically-determined positions of high or maximum probability for hydrogen bond acceptor, donor, acidic and basic groups only has also been published [24].

2.4 3-point versus 4-point pharmacophores

Moving from 3-point pharmacophores to 4-point pharmacophores introduces aspects of 3D shape into the measure, and enables chirality to be resolved. Clearly much more information can be generated; figure 3 shows this for an endothelin antagonist, where there is a much larger number of potential pharmacophores generated with the 4-point definition. Similarity studies (see section 4) indicate that this extra information is meaningful. It appears that the noise tends to remain at a fixed absolute level, whereas compounds expected to be similar are observed to be more similar (in terms of common potential pharmacophores) using 4-point methods. Studies [3, 7] with sets of receptor antagonists, enzyme inhibitors and 'random' compounds showed that the 4-point method could identify proportionately more 'characteristic' pharmacophores for a 7TM-GPCR receptor antagonist set (i.e. pharmacophores that occurred multiple times only in the receptor antagonist set).

2.5 Use of 'relative' pharmacophoric similarity and diversity

The use of this method has been reported [7, 11, 13] where the 'special' feature is a known 'privileged' group for 7-transmembrane G-protein coupled receptor (7TM-GPCR) targets, for the design of combinatorial libraries. Known ligands reported in the MDL Drug Data Report (MDDR) database [25] were analysed, to measure which potential pharmacophores occur more frequently relative to certain 'privileged' substructures. A similar analysis was performed on the company screening set, and a quantifiable target could be defined: to design libraries that exhibit the potential pharmacophores containing the privileged features found in the MDDR set, but which were missing from the company set.

The method can be used for many other applications, both ligand and structure-based. The special feature can be assigned to any motif or group of interest, providing new profiling and design methods. A ligand-based example is to explore the diversity at a specific position for combinatorial library compounds. In this case, the 'special' point will be on the scaffold itself, located at the attachment point of interest. Reagents themselves can be evaluated and diversity profiled based just on their pharmacophoric diversity relative to the attachment point.

A structure-based example is to explore a given subsite in a protein active site. The 'special' feature is positioned as a site point where the molecules to dock will be anchored; this could be done using the position of the atom of the docked scaffold which is used to link the reagents or the derivative substituents. All the potential pharmacophores that include the complementary site points of the sub-site of interest and the special site point are thus calculated and stored. In addition, special coding of the molecules (putative ligands) is required, again assigning the 'special' feature to the point of attachment. It is then possible to compare the pharmacophores calculated for the protein subsite with the ones calculated for the molecules. The results of this analysis could guide the choice of reagents for a library, or indeed point out the fact that none of the reagents (or reagent–scaffold combinations) can explore some of the pharmacophoric 'diversity' exhibited by the site.

2.6 Use of the protein site for steric constraints

An extension of the method that enables the steric shape of a protein site to be used as an additional constraint in the comparison of multiple potential pharmacophores of a protein site with a ligand has been developed, and is being commercialised as the DiR module of Chem-X. The method is equivalent to simultaneous 3D-database searching using multiple 3D pharmacophoric queries and steric constraints; the advantage is that only one conformational sampling is necessary.

3. BCUT CHEMISTRY SPACE – DIVERSESOLUTIONS (DVS)

The DiverseSolutions (DVS) program, developed by Pearlman and Smith [15-19] enables the generation and use of multi-dimensional chemistry-spaces that discriminate the diversity of a large set of compounds. Novel 'BCUT' metrics (descriptors of atomic/molecular properties) are used, which are claimed to have advantages over more traditional descriptors such as molecular weight, ClogP or topological indices, as they reflect both molecular connectivities and atomic properties relevant to intermolecular interactions. 5- or 6-dimensional chemistry-spaces are generally identified for datasets of 50,000 to 1 million compounds. The use of 'bins' for values along each dimension enables a cell-based analysis, with the associated benefits of rapid comparison of different datasets and identification of voids.

A recent modification to the method has been reported [18] that involves the use of a subset of the dimensions for a set of structures with similar biological activity. This subset from an activity-seeded structure-based clustering has been called the 'receptor-relevant' BCUT chemistry-space, and was used to perform a number of validation studies [18, 19]. The method can be considered as a type of 'relative' similarity and diversity, where the subset of properties that are considered to be important for biological activity are separated from the others. This is important, as non-critical properties can be explored, while maintaining the values of the 'receptor relevant' properties.

3.1 Receptor-relevant Sub Chemistry Spaces

Pearlman and Smith have reported the receptor-relevant subspace concept [18], using ACE inhibitors as an example. They found that only 3 out of the 6 'universal' BCUT metrics were 'receptor-relevant', in that the actives clustered only in these 3 dimensions. These 3 metrics are then considered 'receptor-relevant' and worthy of being constrained, while the other can be varied and not 'falsely' constrained.

4. STUDIES USING ABSOLUTE SIMILARITY AND DIVERSITY

4.1 Analysis of reference databases

4.1.1 Multiple 4-point potential pharmacophores

The 3-point and 4-point pharmacophore methods can be used to analyse and compare different sets of compounds and databases. Figure 7 illustrates the 4-point pharmacophores for the MDDR database [25], the Available Chemical Directory (ACD) [25], a company registry database and a set of combinatorial libraries reported by Mason and co-workers [7, 11, 13]. Previous studies [3] had shown the increase in resolution possible using 4-point instead of 3-point pharmacophores.

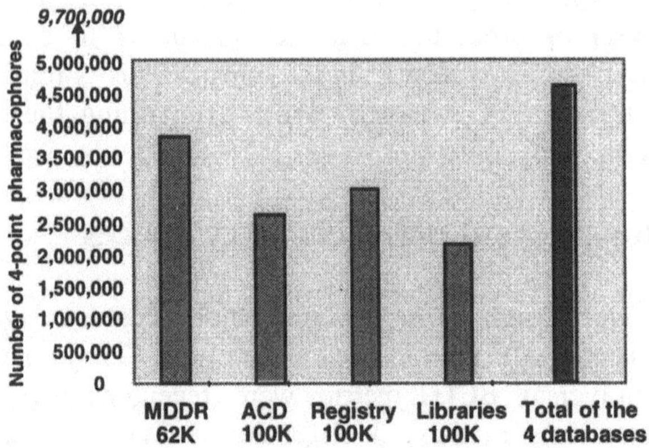

Figure 7. 4-point multiple potential pharmacophores for four reference databases, using 6 features per point and 10 distance ranges.

4.1.2 DVS atomic/molecular properties

Figure 8 illustrates the occupied cells reported [11, 13] for the reference databases using a six dimensional BCUT chemistry space and six bins per axis (charge, electronegativity, H-bond donor, two types of polarizability, molecular surface area) derived using DiverseSolutions.

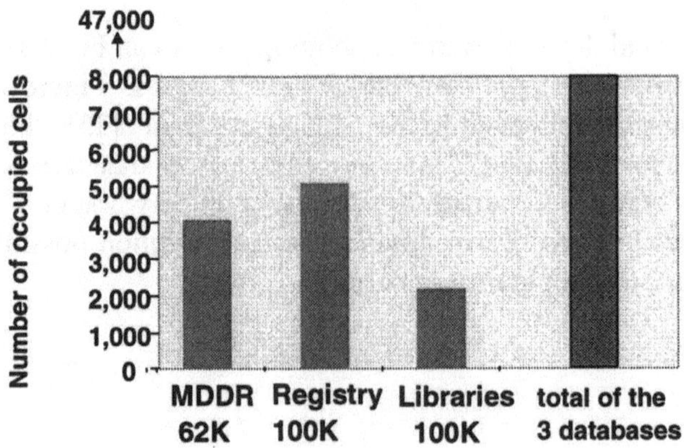

Figure 8. Chemistry space occupation for reference databases, using a 6 dimensional BCUT chemistry space derived using DiverseSolutions and 6 bins per axis.

4.2 Ligand studies

The multiple potential pharmacophore method, used in an absolute or relative sense, provides a powerful new tool for 3D similarity studies. As an example, two endothelin receptor antagonists with about 20 nM activity as antagonists of the ET_A receptor were compared [21]. Figure 9 shows the numbers of potential 4-point pharmacophores and overlapping pharmacophores. The two compounds have very low 2D similarity, but have significant overlap of their 4-point potential pharmacophores, illustrating the power of the method to find similarity between compounds with similar biological activities.

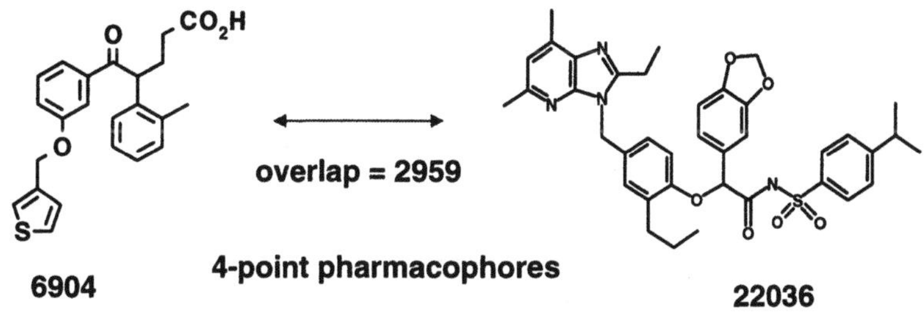

Figure 9. Ligand-ligand 3D similarity: Total and common (overlap) multiple 4-point potential 3-D pharmacophores for two potent endothelin antagonists.

The ability of the pharmacophore method to identify and focus on features important for drug-receptor interactions was important for this result; for example, the assignment of the acidic feature to the acylsulfonamide group increases the overlap by about a third (acids were also considered as general hydrogen-bond acceptors for this analysis).

4.3 Ligand-receptor studies

The multiple potential pharmacophore key calculated from a ligand can be compared to the multiple potential pharmacophore key of complementary site-points in its target binding site. This provides a novel method to measure similarity when comparing ligands to their receptors, with applications such as virtual screening and structure-based combinatorial library design.

An example of the method has been published [14] that compares studies on three closely related serine proteases: thrombin, factor Xa and trypsin. 4-point multiple potential pharmacophore keys were generated from site-points positioned in the active sites using the results of GRID analyses. These are illustrated in figure 10, together with the number of overlapping pharmacophores between the sites; 120 pharmacophores are common to all the protein sites. Figure 11 illustrates the number of overlapping 3-point pharmacophores; there are a similar number in common (121) but many fewer that discriminate between the protein sites.

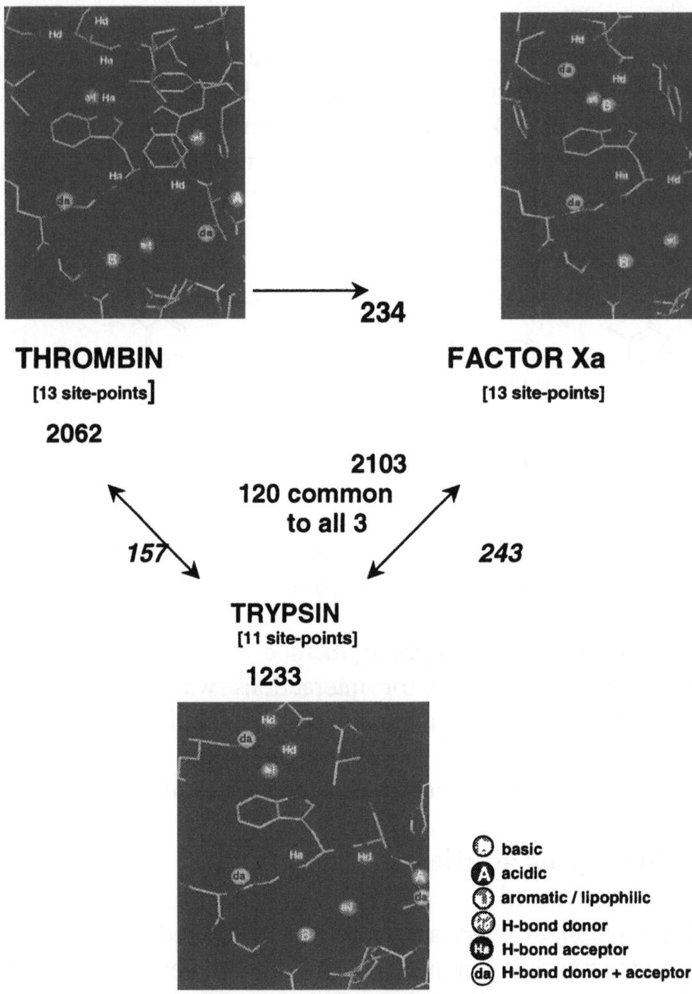

Figure 10. The numbers of potential 4-point pharmacophores calculated on the basis of complementary site-points placed in the active sites of thrombin, factor Xa, and trypsin, and number of overlapping pharmacophores (pair-wise and for all 3 serine protease sites).

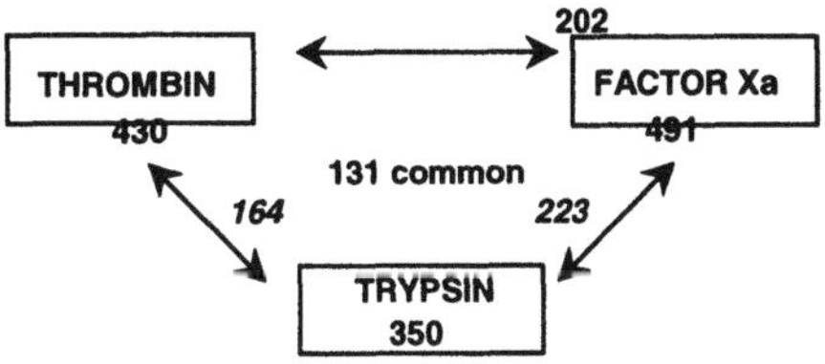

Figure 11. The number of potential 3-point pharmacophores calculated on the basis of complementary site-points placed in the active sites of thrombin, factor Xa, and trypsin, and number of overlapping pharmacophores (pair-wise and for all 3 serine protease sites).

Keys were also generated for some highly selective and potent thrombin and factor Xa inhibitors, using full conformational sampling. Figure 12 shows the number of overlapping pharmacophores between these ligands and the protein active sites. The aim was to investigate whether receptor-based similarity as a function of common potential 4-point pharmacophores for each ligand/receptor pair could replicate the observed enzyme selectivity; the goal was not to predict binding affinities using these overlaps. The expected benefit of using 4-point pharmacophores with their improved shape information was probed by using identical studies with only 3-point pharmacophores.

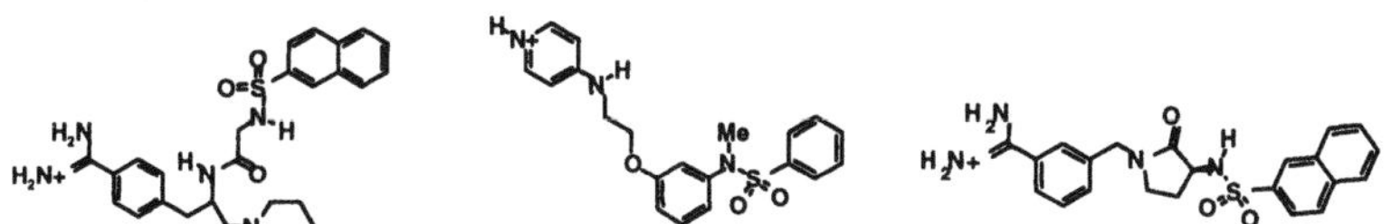

NAPAP Ki 6nM Thrombin		***Boeh. Mann.***[13] Ki 70nM Thrombin		***RPR 118071***[14] Ki 80 nM Factor Xa	
Pharmacophore Overlaps:	**4 pt.**		**4 pt.**		**4 pt.**
Thrombin	**352 ←**	**Thro.**	**135 ←**	**Thro.**	**32**
Factor Xa	**210**	**FXa**	**104**	**FXa**	**59 ←**
Trypsin	**82**	**Tryp.**	**40**	**Tryp.**	**32**

Figure 12. The numbers of overlapping potential 4-point pharmacophores for ligands with those calculated for the active sites of thrombin, factor Xa, and trypsin on the basis of complementary site-points; the arrow points to the enzyme for which the ligand shows biological activity).

The results shown in figure 12 indicate that the use of just 4-point potential pharmacophores can give correct indications as to relative selectivity for ligands for this set of related enzymes. The thrombin and factor Xa inhibitors exhibit greater similarity with the complementary 4-point potential pharmacophore keys of the thrombin and factor Xa active sites, respectively, than with the potential pharmacophore keys generated from the other enzymes; for actual binding energies, other factors such as the strength of hydrogen bonds and hydrophobic interactions will be important.

Figure 13 shows the same overlaps using 3-point potential pharmacophores, and the poorer ability to replicate the observed enzyme selectivity is clear, with two compounds showing maximum overlaps for the wrong enzyme. It would thus appear that the enhanced resolution of the 4-point method is needed for comparisons based just on pharmacophore overlap, without taking any further account of the shape of the site. The new DiR (Design in Receptor) module for Chem-X will use the active site sterics as an additional constraint, giving a method that is equivalent to doing multiple 3D database pharmacophoric searches, using each potential 3- or 4-point pharmacophore as a 3D query, but with only one conformational sampling step per molecule.

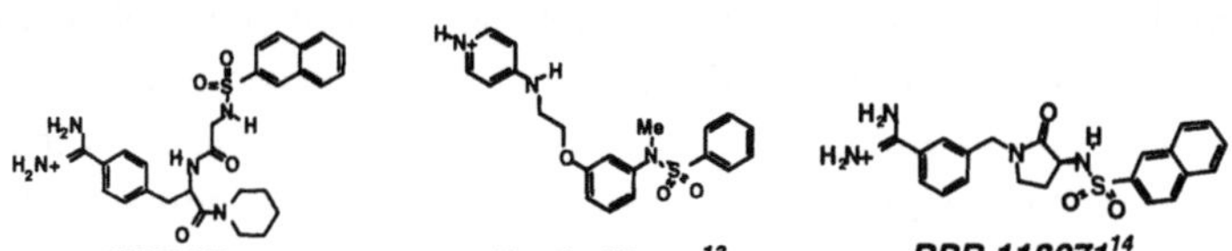

NAPAP Ki 6 nM thrombin		*Boeh. Mann.*[13] Ki 70 nM thrombin		*RPR 118071*[14] Ki 80 nM Factor Xa	
Pharmacophore Overlaps:	3 pt.		3 pt.		3 pt.
Thrombin	168 ←	Thro.	58 ←	Thro.	→ 79
Factor Xa	144	FXa	→ 64	FXa	56 ←
Trypsin	99	Tryp.	31	Tryp.	49

Figure 13. The number of potential 3-point pharmacophores for ligands that overlap with those calculated for the active sites of thrombin, factor Xa, and trypsin on the basis of complementary site-points; the right-side arrow and box points to the enzyme for which the ligand shows binding.

To evaluate this method for ligand-receptor similarity in the context of compound design and virtual screening, the above analysis was repeated using two fibrinogen receptor antagonists (see figure 14). These compounds have 2D structural features (e.g. benzamidine) that resemble trypsin-like serine protease inhibitors, but had no reported activity for this class of enzymes. With 4-point pharmacophore profiling, the degree of similarity is very small, whereas using 3-point pharmacophores, the molecules exhibited some pharmacophoric similarity against all three enzymes, with MDDR 192259 showing significant overlap.

Fibrinogen receptor antagonists

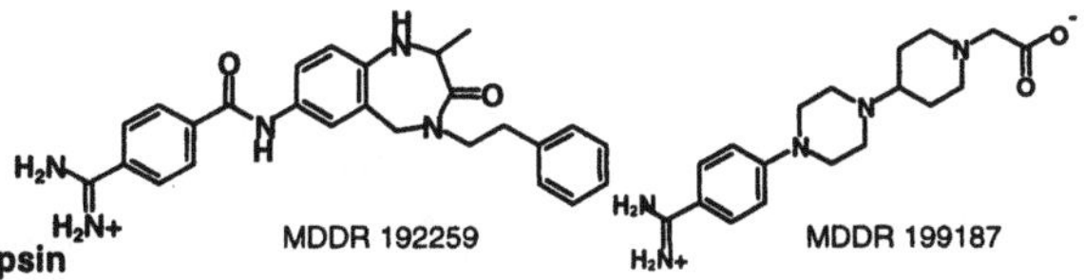

Thrombin / Factor Xa / Trypsin	MDDR 192259			MDDR 199187		
4-point pharmacophore overlap	**4**	**2**	**0**	**0**	**0**	**0**
3-point pharmacophore overlap	**60**	**57**	**48**	**3**	**5**	**1**

Figure 14. The number of common potential 4- and 3-point pharmacophores for 'inactive' ligands with similar 2D structural motifs to active compounds with those calculated from the complementary site-points for the active sites of thrombin, factor Xa, and trypsin.

5. STUDIES USING RELATIVE SIMILARITY AND DIVERSITY

The use of absolute similarity and diversity methods provides useful information about overall similarities and differences between molecules, or databases of molecules, but for certain similarity and diversity design applications, the use of a relative measure that is internally referenced to a feature or substructure of interest can provide a more powerful tool. This is true for a receptor-based design and for ligand-based design around a feature or substructure of interest. A common frame of reference enables ligand-based and receptor-based studies to be combined, for example with reagents in an active site and on a template

5.1 Ligand – receptor studies using multiple potential pharmacophores

The ligand-receptor similarity studies can be further enhanced by applying the 'relative' similarity concept. Thus for the serine protease/ligand studies, the pharmacophores could be focused around a point in the S1 recognition pocket. Figure 15 illustrates this for the MQPA ligand, which using normal 3- or 4-point multiple potential pharmacophore comparisons has ambiguous or incorrect selectivity, having more or equal overlapping potential pharmacophores with the wrong enzyme (factor Xa instead of thrombin); the enhanced resolution is clearly seen using only potential pharmacophores that contain the S1 basic point.

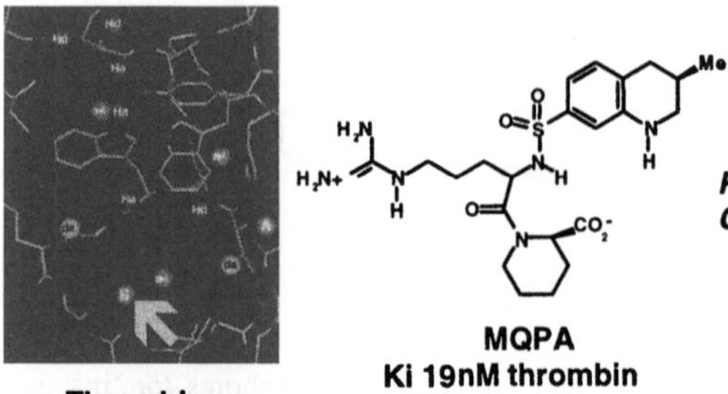

Pharmacophore Overlaps:		3 pt.	4 pt.	4pt. relative	
Thrombin:		183	192	76	←
Factor Xa:	→	199	191	50	
Trypsin:		128	94	34	

Figure 15. 3- and 4-point multiple pharmacophore overlaps for the thrombin ligand MQPA and the serine protease active-site derived pharmacophores; the left-side arrow indicates the incorrect indication of factor Xa selectivity from the 3-point figures, and the right-side arrow the observed activity and the increased resolution of selectivity using the 4-point relative pharmacophores.

5.2 Library design using multiple potential pharmacophores

The use of a 'relative' method of pharmacophoric diversity has been reported [7, 11-14] where the 'special' feature is assigned to various known 'privileged' groups for 7TM-GPCR targets. The goal was the design of combinatorial libraries that would enrich the existing screening set. An analysis was made of the MDDR drugs database to identify known ligands, and these were pharmacophorically analysed, so as to measure which potential pharmacophores exist relative to the 'privileged' substructures. Libraries were then designed to exhibit potential pharmacophores relative to the privileged substructures to fill diversity voids, defined as those potential pharmacophores that were found in the MDDR set, but were missing from the company set.

5.3 Analysis of active compounds using DVS

The example reported by Pearlman and co-workers [18, 19, 26] involves the analysis of ACE inhibitors. They found that 3 of the 6 BCUT metrics they had identified from an analysis of the MDDR database (60,000 'drug' compounds) were 'receptor relevant', i.e. that the actives clustered in these dimensions. Figure 16 shows the actives 'clustering' in a sea of general 'drugs' (MDRR compounds) in these 3 dimensions.

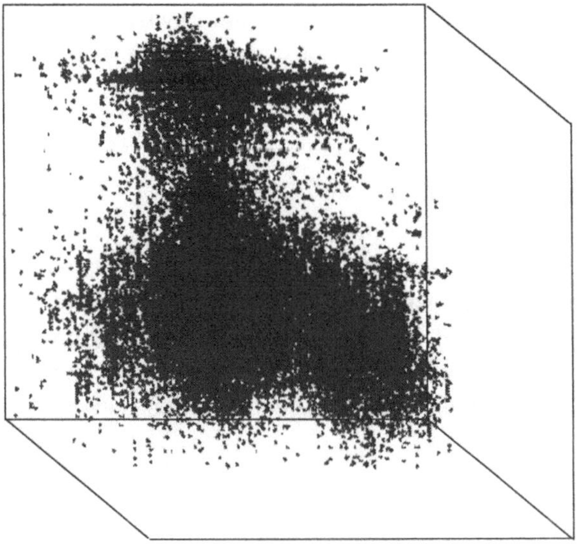

Figure 16. A plot of ACE inhibitors (black) clustering within a 'receptor-relevant" subset of the MDRR (grey). The axes are the 3 receptor-relevant descriptors.

Figure 17 illustrates how the same compounds look with a 'non-relevant' descriptor substituted in; there is clearly a wide range of values and it would be too restrictive to fix a particular value.

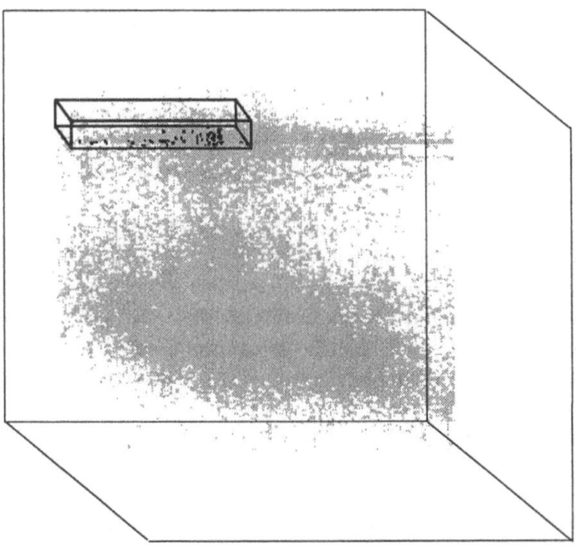

Figure 17. A plot of ACE inhibitors (black) with a 'receptor-relevant" subset of the MDRR (grey). The axes are 2 receptor-relevant descriptors and a non-relevant metric (number 5).

Interestingly, the ACE-receptor-relevant metrics chosen as being 'receptor-relevant' appear to be consistent with the published binding model.

6. CONCLUSIONS

The use of 'relative' similarity and diversity methods can add powerful new methods for design and analysis. The multiple potential pharmacophore method has been described, and its application to practical design problems discussed. These studies highlight the importance of 4-point pharmacophores and the use of special centres to focus diversity studies.

ACKNOWLEDGMENTS

The author would like to thank Daniel Cheney at Bristol-Myers Squibb and colleagues at Rhone-Poulenc Rorer, in particular Paul Menard, Isabelle Morize and Richard Labaudiniere.

REFERENCES

1. Pickett, S.D., Mason, J.S. and McLay, I.M. Diversity Profiling and Design Using 3D Pharmacophores: Pharmacophore-Derived Queries (PDQ). *J. Chem. Inf. Comput. Sci.,* 1996, **36**, 1214-1223.
2. Ashton, M.J., Jaye, M.C. and Mason, J.S. New Perspectives in Lead Generation II: Evaluating Molecular Diversity. *Drug Discovery Today,* 1996, **1**, 71-78.
3. Mason, J.S. and Pickett, S.D. Partition-based selection. *Perspect. Drug Disc. Des.,* 1997, **7/8**, 85-114.
4. Davies, E.K. and Briant, C. Combinatorial Chemistry Library Design Using Pharmacophore Diversity. Accessible through URL: http://www.awod.com/netsci/Science/Combichem/feature05.html.
5. Davies, K. Using pharmacophore diversity to select molecules to test from commercial catalogues. In *Molecular Diversity and Combinatorial Chemistry. Libraries and Drug Discovery*, Eds. Chaiken, I.M. and Janda, K.D., 1996, Washington: American Chemical Society, pp. 309-316.
6. Mason, J.S. Pharmacophore similarity and diversity: Discovery of novel leads for cardiovascular targets. In *Lead Generation and Optimization*, Strategic Research Institute, New York, 1996 (March 21-22, New Orleans meeting).
7. Mason, J.S., Morize, I, Menard, P.R., Cheney, D.L., Hulme, C. and. Labaudiniere, R.F. A new 4-point pharmacophore method for molecular similarity and diversity applications: Overview of the method and applications, including a novel approach to the design of combinatorial libraries containing privileged substructures. *J. Med. Chem.*, submitted for publication.

8. Chemical Design Ltd, part of OMG, Oxford Science Park, Oxford OX4 4GA, UK.
9. Murrall, N.W. and Davies, E.K. Conformational freedom in 3D databases. 1. Techniques. *J. Chem. Inf. Comput. Sci.*, 1990, **30**, 312-316.
10. Evans, B.E., Rittle, K.E., Bock, M.G., DiPardo, R.M., Freidinger, R.M., Whitter, W.L., Lundell, G.F., Veber, D.F., Anderson, P.S., Chang, R.S., Lotti, V.J., Cerino D.J., Chen, T.B., Kling P.J., Kunkel, K.A., Springer, J.P. and Hirshfield J. Methods for Drug Discovery: Development of potent, selective, orally effective cholecystokinin antagonists. *J. Med. Chem.*, 1988, **31**, 2235-2246.
11. Mason, J.S. Diversity for drug design: A multiple-technique approach. In *Exploiting Molecular Diversity*, Proceedings of CHI Meeting, Coronado, CA, March 2-4 1998.
12. Mason, J.S. and Cheney, D.L., Absolute and relative diversity/similarity approaches using both ligand and protein-target-based information. In *Chemoinformatics*, Proceedings of CHI Meeting, Boston, MA, June 15-16, 1998.
13. Mason, J.S. and Cheney, D.L. Recent advances in pharmacophore similarity in structure-based drug design. In *Book of Abstracts, 215th ACS National Meeting*, Dallas, March 29-April 2, 1998, American Chemical Society, Washington, D. C.
14. Mason, J.S. and Cheney, D.L. Ligand-receptor 3-D similarity studies using multiple 4-point pharmacophores. In *Biocomputing*, Proceedings of the 1998 Pacific Symposium, World Scientific Publishing Co. Pte. Ltd., Singapore, 1999, pp 456-467.
15. *DiverseSolutions* was developed by R.S.Pearlman and K.M Smith at the University of Texas, Austin, TX and is distributed by Tripos Inc, St. Louis, MO
16. Pearlman, R.S. *DiverseSolutions User's Manual*, University of Texas, Austin, TX, 1995.
17. Pearlman, R.S. Novel Software tools for Addressing Chemical Diversity, *Network Science*, 1996, http://www.awod.com/netsci/Science/combichem/feature08.html
18. Pearlman, R.S. and Smith, K.M. Novel software tools for chemical diversity, *Perspect. Drug Disc. Des.*, 1998, **9**, 339-353.
19. Pearlman, R.S. and Smith, K.M. Metric Validation and the receptor-relevant subspace concept. *J. Chem. Inf. Comput. Sci.*, 1999, **39**, 28-35.
20. Mason, J.S. Experiences with Searching for Molecular Similarity in Conformationally Flexible 3D Databases. In *Molecular Similarity in Drug Design*, Ed. Dean, P. M., Blackie Academic and Professional, Glasgow, 1995, pp. 138-162.
21. Astles, P.C., Brealey, C., Brown, T.J., Facchini, V., Handscombe, C., Harris, N.V., McCarthy, C., McLay, I.M., Porter, B., Roach, A.G., Sargent, C., Smith, C. and Walsh, R.J.A. Selective endothelin A receptor antagonists. 3. Discovery and structure-activity relationships of a series of 4-phenoxybutanoic acid derivatives. *J. Med. Chem.*, 1998, **41**, 2732-2744.
22. Balducci, R., McGarity, C., Rusinko III, A., Skell, J., Smith, K. and Pearlman, R.S. Laboratory for Molecular Graphics and Theoretical Modeling, College of Pharmacy, University of Texas at Austin; distributed by Tripos Inc.: 1699 S. Hanley Road, Suite 303, St. Louis, MO 6314.
23. Molecular Discovery Limited, West Way House, Elms Parade, Oxford OX2 9LL, England
24. Mills, J.E.J. and Dean, P.M. Three-dimensional hydrogen-bond geometry and probability information from a crystal survey. *J. Comput.-Aided Mol. Des.*, 1996, **10**, 607-622.
25. MDL Information Systems Inc., 14600 Catalina Street, San Leandro, CA 94577, USA.
26. Pearlman, R.S. and Deandra, F. *Manuscript in preparation.*

Chapter 5

Diversity in Very Large Libraries

Diversity in Very Large Libraries

Lutz Weber and Michael Almstetter
Morphochem AG, Am Klopferspitz 19, 82152 Martinsried, Germany

Key words: Combinatorial chemistry, genetic algorithms, combinatorial optimisation, QSAR, evolutionary chemistry, very large compound libraries

Abstract: Combinatorial chemistry methods can be used, in principle, for the synthesis of very large compound libraries. However, these very large libraries are so large that the enumeration of all individual members of a library may not be practicable. We discuss here how one may increase the chances of finding compounds with desired properties from very large libraries by using combinatorial optimisation methods. Neuronal networks, evolutionary programming and especially genetic algorithms are heuristic optimisation methods that can be used implicitly to discover the relation between the structure of molecules and their properties. Genetic algorithms are derived from principles that are used by nature to find optimal solutions. Genetic algorithms have now been adapted and applied with success to problems in combinatorial chemistry. The optimisation behaviour of genetic algorithms was investigated using a library of molecules with known biological activities. From these studies, one can derive methods to estimate the diversity and structure property relationships without the need to enumerate and calculate the properties of the whole search space of these very large libraries.

1. INTRODUCTION

In nature, the evolution of molecules with desired properties may be regarded as a combinatorial optimisation strategy to find solutions in a search space of unlimited size and diversity. Thus, the number of all possible, different proteins comprising only 200 amino acids is 20^{200}, a number that is much larger than the number of particles in the universe (estimated to be in the range of 10^{88}). Similarly, the number of different molecules that could be synthesised by combinatorial chemistry methods far exceeds our synthetic and even computational capabilities in reality. Whilst diversity and various properties of compound libraries in the range of several thousands to millions can be calculated by using a range of different methods, there is little available knowledge and experience for dealing with very large libraries. The task for chemists is therefore to find methods that can be used to choose useful subsets from this practically unlimited space of possible solutions.

The intellectual concept and the emerging synthetic power of combinatorial chemistry are moving the attention of experimental chemists towards a more abstract understanding of their science: instead of synthesising and investigating just a few molecules they are dealing now with libraries and group properties. The answers to questions such as how diverse or similar are any two compounds, are now not just intellectually interesting but also have commercial value. Therefore, the ability to understand and use very large libraries is, in our opinion, connected to the understanding and the development of chemistry in the future.

The discovery of a new medicine may be understood as an evolutionary process that starts with an initial knowledge set, elaborating a hypothesis, making experiments and thereby expanding our knowledge. A new refined hypothesis will give rise to further cycles of knowledge and experiments ending with molecules that satisfy our criteria. If very large compound libraries are considered, one may argue that the desired molecules are already contained within this initial library. A very large library on the other hand means that we are neither practically nor theoretically able to synthesise or compute all members of this library. How can we nevertheless find this molecule? Is it possible to develop methods that automate the discovery of new medicines by using such libraries without human interference?

An answer to these questions would be a novel approach to combinatorial chemistry that tries to connect the selection and synthesis of biologically active compounds from the very large library by mathematical optimisation methods. Heuristic algorithms, like genetic algorithms or neural networks, mimic the Darwinian evolution and do not require the *a priori* knowledge of structure-activity relationships. These combinatorial optimisation methods [1] have proved to be useful in solving multidimensional problems and are now being used with success in various areas of combinatorial chemistry. Thus, evolutionary chemistry may aid in the selection of information rich subsets of available compound libraries or in designing screening libraries and new compounds to be synthesised, adding thereby a new quality to combinatorial chemistry.

2. GENETICS OF MOLECULES

Natural and artificial evolutionary systems are composed of at least two layers - encoding (genotypes) and realisation (phenotypes). Both layers are connected by an operator providing the recipe for constructing the phenotype from the genotype. The principles of evolution can be generalised in a simple scheme that displays the basic elements for the implementation of artificial evolutionary systems (figure 1). Generally, new generations of genotypes with better fitness are evolved based on the fitness of the phenotypes of previous generations in a feedback loop.

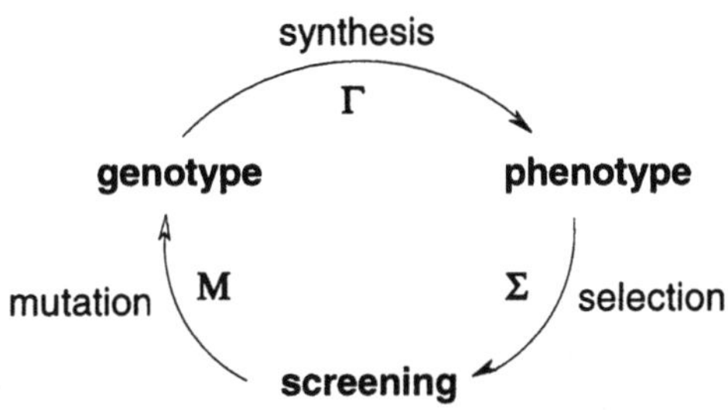

Figure 1. A generalised representation of an evolutionary system Genotypes and phenotypes are connected by an operator (generator **Γ**) that generates the phenotypes from their genotypes. A selector (**Σ**) provides feedback about the fitness of phenotypes to the level of genotypes. New genotypes are then generated according to the fitness of the phenotypes of the first generation of genotypes by mutation (**M**).

It is important to note for later understanding that the genotype, e.g. the triplet UUC in nature's encoding scheme for the amino acid phenylalanine, does not reflect any physico-chemical property of the encoded phenotype directly.

Thus, the first step in using evolutionary algorithms for all kinds of molecules is to invent a suitable encoding system for the chemical space of interest. One of the most important inventions for structure-based chemistry was the Van't Hoff valence-bond-based structural description of molecules in the last century. A general structure-based algebraic representation of molecules has been developed first by Ugi [2] using *be*- and *r*-matrices. Within this representation all atoms of a molecule, their connectivity and shared electrons are used for formal metric "chemical distance" between isomeric molecules [3]. From this concept, a formal description of a "universal" structure-based diversity measure can be derived when counting changes in these *be*-matrices that are needed to generate one molecule from another one. More recently, Weininger [4] has introduced the elegant SMILES notation for molecules. Both methods allow the unambiguous reconstruction of a Van't Hoff type structure from this encoding according to a defined set of rules. These rules can be encoded themselves and used to build whole molecules [5, 6] in a more comprehensive way using e.g. >C<, -H, -O-, >C=O as building blocks.

Alternatively, combinatorial libraries allow an efficient DNA-like encoding using arbitrary binary bit-strings [7, 8] or decimal bit-strings [9] as depicted (figure 2). As in nature, an operator **Γ** is needed that translates this encoding into a general valence bond type representation or into real molecules, respectively. **Γ** may be understood also as a synthesis scheme or a routine that drives an automated synthesiser generating this particular molecule.

smiles:

NC(=O)CN <u>(C(C)C)</u> **C(=O)CN** (Cc1ccc(O)cc1) **C(=O)CN** ***2CCCCC2***

decimal:

1 <u>2</u> **1** 88 **1** ***101***

binary:

<u>1001</u> 1110 ***1111***

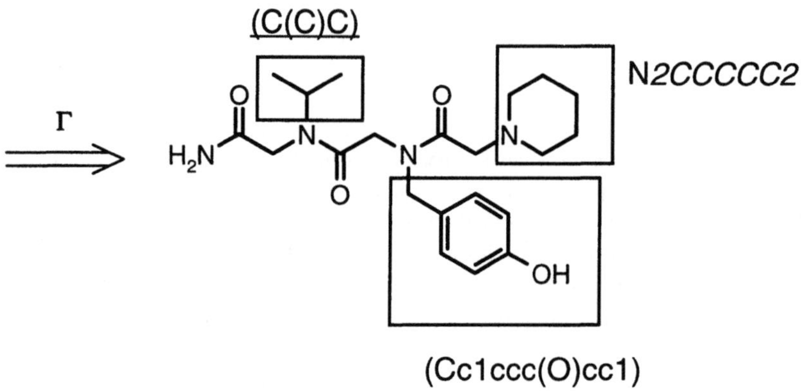

Figure 2. Encoding of combinatorial libraries: the example shown is a tripeptoid type structure that allows various encoding schemes needed for evolutionary algorithms. Corresponding sub-structures and encoding schemes are marked by the same typography.

3. IMPLEMENTING ARTIFICIAL EVOLUTION

A large number of various mathematical algorithms have been developed to select optimal combinations from a pool of combinatorial possibilities. Most appealing are certainly the evolutionary algorithms developed by Rechenberg [10] at the Technical University of Berlin. These algorithms were applied to the calculation of optimal wing profiles of airplanes. In his basic work, a wide variety of different **Σ** and **M** operators have been defined giving raise to a series of possible optimisation strategies. One special sub-case of evolutionary optimisation algorithms are the "genetic algorithms" (GAs), which were developed by Holland [11]. The name is due to the similarity with the principles of DNA-based evolution in living organisms. The number of applications of evolutionary and especially genetic algorithms to problems in chemistry is small but increasing rapidly, as shown in recent overviews [12, 13].

A series of other optimisation algorithms include fuzzy logic, neuronal networks, cellular automata, fractals and chaos theory and mixed variants that will not be covered in this chapter and have been termed "soft computing" methods [14]. Compared to all other algorithms, GA's are ideal for optimisation problems where the search space is very large and the solution consists of a subset of a large number of parameters each of which can be independently set. GA's are stochastic and each run may provide a different result — providing rather a set of good solutions instead of the best solution. These properties make GA's ideal candidates for dealing with large combinatorial libraries where the independent parameters are either the large number of starting materials for synthesis, or alternatively the substructures found in the complete molecule.

3.1 Operators of Genetic Algorithms

3.1.1 Operator M acting on the genome

A genetic algorithm usually starts with a randomly chosen set of **n** different entities encoded by genomes - the population. The evaluation and ranking of the fitness of these genomes is performed according to the fitness of their phenotypes in the selection step **Σ**. Various methods are possible for selecting parents that may create the new generation of offspring — the "children". These new genomes are then generated from this ranked list by the **M** GA functions that are inspired by those of DNA like genetics: *death*, *replication*, *mutation*, *deletion*, *insertion* and *crossover* (figure 3).

Replication regenerates an equivalent individual (child = parent). *Mutation* sets one or more bits in the parent genome to a different value to obtain a new child. *Crossover* takes two or more genomes to build new genomes by mixing them according to various rules. *Deletion* deletes a bit, or bit strings, from the parent genome, *insertion* introduces new bits or bit strings. According to the chosen encoding scheme, these functions may have a different meaning or sense when applied to the genome. Thus, crossover applied to a genome with a binary representation may intersect not only between different substituents as it is only possible with decimal encoding, but also "in-between" a substituent (see figure 3). While the first strategy of replacing only whole building blocks appeals more to the idea of chemists, the latter is more similar to crossover in DNA, corresponding thereby to the technique used by nature that may be regarded as a mixture of a joint mutation and crossover. Contrary to evolution in nature, we are completely free to define how these GA operators are applied in the computer, e.g. a new child may have more then just one or two parents.

Death:		1001 1110 1111	=>	
Replication:		1001 1110 1111	=>	1001 1110 1111
Mutation:		1001 1110 1111	=>	1**1**01 1110 1111
Deletion:		1001 1110 1111	=>	1001 110 1111
Insertion:		1001 1110 1111	=>	1001 111**01**0 1111
Crossover:	parent 1	**1001 1**110 1111	=>	**1001 1001 1001**
	parent 2	1001 0**001 1001**	=>	1001 0110 1111

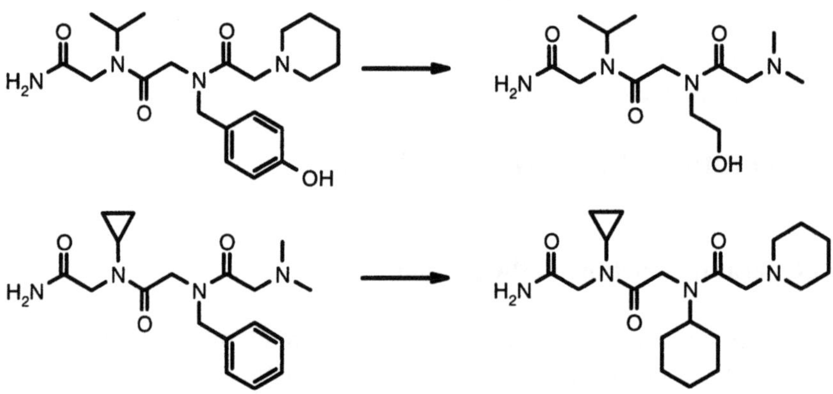

Figure 3. Operators of genetic algorithms as they may be applied to the encoding of molecules from combinatorial libraries. The changes in the bit-strings are indicated by the use of bold-face type. The tripeptoid example from figure 2 has been chosen to illustrate the DNA-like crossover with binary bit strings

3.1.2 Operator Σ for ranking and selection

After evaluating the fitness of the molecules, the genetic algorithm includes a ranking and selection step **Σ** that determines which genome is subject to which **M** operator. This step is a more or less strong implementation of the idea of "survival of the fittest" where fit genomes are allowed to survive and weak genomes are eliminated. The methods that are found in the literature differ significantly for **Σ**. First, a ranked list of **n** genomes is generated. Second, the genomes that are subject to death are determined, e.g. genomes that are older than a specified number of cycles. The remaining list of a predetermined population size is then treated by combinations of **M** either with equal, stochastic, or a distribution according to the rank of the genome in this list. Thus, in an example method called "best third", the worst third of all genomes is eliminated, the best third of the genomes are simply replicated to the new generation, the middle third is subject to mutation and crossover to generate new children (figure 4) giving raise to a "elitist" selection.

genomes	rank		new genome
1001 1110 1111	1	=>	1001 1110 1111
1001 1110 1111	2	=>	1**1**01 1110 1111
1001 1110 1111	3	=>	**1001 1001 1001 + 1001** 0**001 1001**
1001 0**001 1001**	4	=>	1001 0110 1111 + 1001 0110 1011
1001 1110 1111	5	=>	-
1001 1110 1111	6	=>	-

Figure 4. Generation of new generation of genomes for evolutionary optimisation with a variant of the "best third method". The changes are indicated in bold-face type.

However, while various combinations and variants are possible, it has not been shown that a specific version is superior. We have recently used a method that stores all results that were obtained during the course of the GA in a database. For the generation of the **n** new offspring, the **n** best genomes are then retrieved from this database. This procedure corresponds to a "best half" method with a flexible treatment of the death parameter, since the **n** fittest genomes are always present and the new children are offspring of these genomes irrespective of their age.

3.1.3 What are optimal GA parameters?

Many parameters can be set and varied during the course of a genetic algorithm experiment: e.g. size of populations, number of surviving genes, mutation rate, number of parents for a number of children and finally the ranking function. This parameter set constitutes the "breeding" recipe for molecules of higher fitness. Finding optimal parameters for a given problem is an optimisation problem by itself. The "structure" of the search space has also a large influence on whether or not a genetic algorithm will be successful [10,15]. We have investigated the influence of population size, mutation rate, encoding and crossover strategy using several biological data sets of a combinatorial library. For a new book on the theory of GA's see [16].

The good search power of GA's is believed to originate from the building-block hypothesis [16-18]. This hypothesis states that the combination of "fit" building blocks or contiguous bit strings or schemes of genes on the genomes may yield higher-order schemes of even better fitness. This optimisation behaviour matches perfectly the discontinuous, non-steady structure space of chemistry that is formed by building blocks that are then statistically analysed by the GA. Such building blocks are e.g. atoms, reagents, starting materials and reactions: large combinatorial libraries are generated from systematic arrays of building blocks.

The implied assumption is that one should obtain a systematic structure-property relationship as well. The task of an optimisation procedure is then to discover, with a low sampling rate, the system that allows one to predict properties of new molecules.

3.2 Computational Methods to Select Similar Compounds

Genetic algorithms have been developed to select molecules from a large virtual library exhibiting structural similarities to a given target molecule, e.g. a known drug. Tripeptoid like molecules have been built [9] in the computer by a genetic algorithm choosing from a set of 2507*2507*3312 preselected building blocks giving a library size of about 20 billion. As a proof of concept and to study the GA's optimisation behaviour in finding the optimum in this combinatorial search space, a specific tripeptoid (figure 2) was chosen out of this library as the target molecule. The similarity of newly generated molecules with this target then became the selection criterion for fitness. A topological descriptor using atom pairs separated by a specific number of bonds was used as a similarity measure. Several GA **Σ** and **M** strategies were studied like the stochastic+best third selection and random+neighbours mutation. In the stochastic selection procedure, parents are chosen randomly from the previous population to generate new children, whereas in the best third method, the top-scoring best third of all parents are transferred unchanged, the worst third is eliminated and the medium third is used to generate new children via crossover. Random mutation permits each gene to be mutated with equal probability, whereas neighbours mutation follows a given rule that a mutation may lead only to a similar building block. Using a population size of 300 molecules, the elitist best third selection and neighbours mutation, the right answer was found in the described peptoid example after only 12 generations! This result is rather astonishing since only 3600 peptoids were examined out of the 20 billion. Known CCK and ACE antagonists then were chosen as molecular targets to search for similar tripeptoids. A striking structural similarity between the proposed peptoids and the target molecules was achieved in this "in silico" experiment generally after only 20 generations.

A genetic algorithm has been used in a similar way to propose new polymer molecules that mimic a given target polymer [5]. The molecules were built computationally by linking predefined main chain or side chain atom groups (the 'genes' of the molecules) together and filtering the products with several chemical rules about stability. Some new interesting GA operators were introduced such as insertion and deletion of genes into chromosomes, shifting main-chain-atom groups into other positions of the chromosome, or blending parent chromosomes into one large chromosome. Even more chemical rules are needed when generating general, non-polymeric molecules of all structural classes with a genetic algorithm [6]. Target molecules with a given molecular weight and a 3-dimensional shape were chosen as an example. The method was stated to be of use for any molecular target function like enzyme inhibitors, polymers or new materials.

An interesting example for selecting active compounds from a large database of general molecules with a GA was presented by Gobbi [19]. Encoding of molecules was performed with a bit string of length 16384 where individual bits are set according to the occurrence of substructural elements. After crossover, the molecule that is most similar to the new offspring was retrieved from the database and added to the population. Once a parent was used more than 10 times, it was eliminated from the parent set. The GA was tested in a simulated experiment against a data set from the National Cancer Institute comprising 19596 biological activities. Using a population size of 20 compounds, most or all highly active compounds were found after examining 1 to 10% of the complete database. This method may replace therefore conventional, "blind" high throughput screening in the future since it allows one to reduce screening costs significantly.

3.3 GA Driven Evolutionary Chemistry

The evolutionary mechanisms of nature have been used for phage display libraries, combinatorial biochemistry or even artificial evolution of enzymes [20]. The idea of the application of evolutionary chemistry to small, non-oligomeric molecules is based on the idea of replacing DNA by genetic algorithms and encoding molecules from a combinatorial library in the computer. Examples have been reported on the successful integration of genetic algorithms, organic synthesis and biological testing in evolutionary feedback cycles.

A population of 24 compounds was randomly chosen from the 64 million possible hexapeptides and optimised for trypsin inhibition using a genetic algorithm [21]. Biological testing was performed with a chromogenic assay with trypsin. According to the previously described "best third" method, the best 6 peptides out of 24 where duplicated, the worst 6 were eliminated and the rest was kept to arrive at a new population of 24 peptides. This new population of peptides was then changed by a crossover rate of 100% choosing 2 parents at random. Thereafter, mutation was applied with a probability of 3% - providing a GA with a slight elitism. The average inhibitory activity was improved from 16% of the first randomly chosen population to 50% in the sixth generation. Moreover, 13 peptides out of the 25 most active peptides comprised a consensus sequence of Ac-XXXXKI-NH_2, eight of which had a Ac-XXKIKI-NH_2 sequence. The best identified peptide was Ac-TTKIFT-NH_2 with an inhibition of 89%, being identical with a previously found trypsin inhibitor from a phage display library.

In another example, only 300 peptides were synthesised in 5 generations to obtain substrates for stromelysin out of the pool of the possible hexapeptides [22]. The peptides were synthesised on solid support and fluorescence marked at the N-terminus. After proteolytic cleavage the non-fluorescent beads were analysed. The starting sequence was biased towards using 60 peptides of the sequence $X_1PX_3X_4X_5X_6$, removing the bias in all subsequent generations. From the populations of 60 peptides, the best was copied to the new generation, the others were changed by a crossover rate of 60%. The new peptides were then subjected to mutation with a rate of 0.1% applied to each bit of the 30 bit gene, giving a 3% overall mutation rate. The GA was terminated when 95% of the population members were identical. The hexapeptide MPQYLK was identified as the best substrate for stromelysin in the final generation, being cleaved between the amino acids tyrosine and lysine. The selectivity of the new substrates versus collagenase was also determined and selective substrate sequences had been identified for both enzymes. Therefore, this method may not only help to find new substrates but to also obtain structure-activity and selectivity ideas for new inhibitors.

The first example of a GA driven selection and synthesis of non-peptidic molecules has been published for the selection of thrombin inhibitors [13].

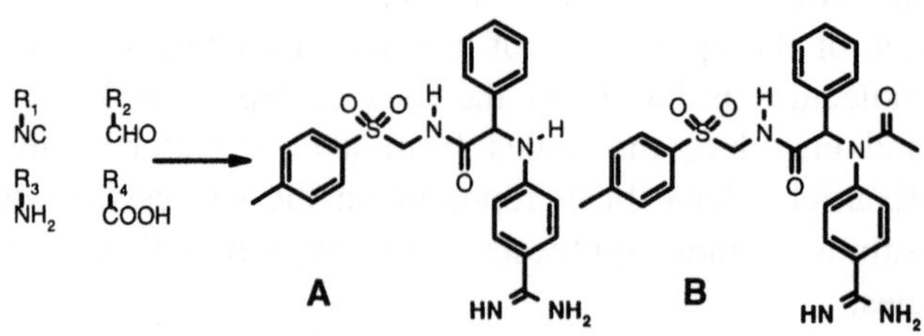

Figure 5. N-aryl-phenylglycine amide type thrombin inhibitor was selected from 160000 possible reaction products with a GA and a thrombin inhibitor assay as the feedback function.

Using 10 isonitriles, 40 aldehydes, 10 amines and 40 carboxylic acids 160000 Ugi four component reaction combinations are possible. In the initial population, the best reaction product exhibited an IC50 of about 300 µM. After 20 generations of 20 single Ugi products per population, a thrombin inhibitor with a sub-micromolar IC50 was discovered. To our surprise, this N-aryl-phenylglycine amide derivative **A** (figure 5) is the three component side product of the four component reaction **B**. However, the encoding was done for the process of combining the four starting materials and not for the final, expected products. The applied GA is obviously not product structure-based and the feedback function depends on the process including also possibly varying yields of the reaction, mistakes in biological screening and so on. A GA is therefore rather tolerant to experimental mistakes and may still yield good results, even if the starting hypothesis is wrong, since false negative results are simply eliminated and not remembered. The elimination of false positive results takes somewhat longer - depending on how often a good genome is allowed to replicate. This "fuzzy" and robust optimisation property makes GA's especially attractive for real time experimental optimisations as described above.

The disadvantage of GA's for general applications is its intrinsic sequential nature since learning takes place only over a rather unpredictable number of optimisation cycles. The speed of a GA driven optimisation strongly depends therefore on the cycle time of synthesis and screening which prevents using long synthesis procedures for compound generation.

3.4 SIMULATED MOLECULAR EVOLUTION

Genetic algorithms are stochastic optimisation procedures lacking a clear theory to guide the design and parameterisation [16]. Optimal GA's have to be developed using real experimental data with trial and error. A first example for the application of GA's for compound selection from large databases has been given by Gobbi [19] for general compound libraries.

We have recently used a combinatorial library of 15360 Ugi three component products to study the optimisation behaviour of GA's [20]. The biological activity of all products of this library was measured against thrombin. We chose 16 isonitriles (C1 - C16), 80 aldehydes (A1 - A80) and 12 amines (B1 - B12) to give a combinatorial library of 16*80*12 = 15360 compounds. Whereas the isonitriles and aldehydes were selected on their availability and coverage of a broad range of diversity, the amines were chosen to fill the thrombin arginine-binding P1 pocket (figure 6). Some of these amines are already known for their affinity to thrombin in the high micromolar range. Due to this structural bias we can expect the molecules of this library to cover a broad range of affinities to thrombin.

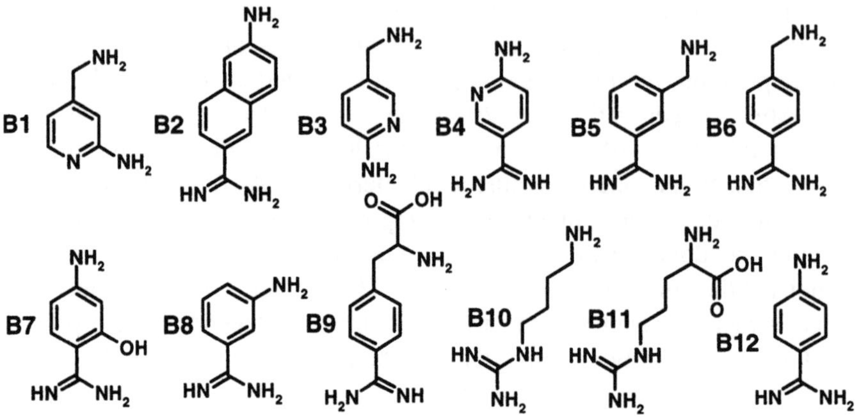

Figure 6. Amines that were used for the 15360 member thrombin inhibitor library.

All final products were evaluated with mass spectrometry to control the quality of the products. The combined data give a structure-activity landscape that provides a model for large databases. On this landscape an "artificial" evolution can be applied *a posteriori* as opposed to the above evolutionary chemistry experiment. The 4 dimensional search space is given by the building blocks A, B and C and the IC50 values for assessing and optimising the performance of a GA in finding the best products. Out of all 15360 products, only 9 (0.059%) exhibited IC50 values below 1 μM. 54 (0.352%) were between 1 and 10 μM and 675 (4.395%) between 10 and 100 μM. The fraction of active products in the library was rather low (<5%) despite the biased nature of our choice of starting materials.

Two implementations of genomes, a binary and a decimal (figure 2), were compared to study the dependence on a DNA-like and chemist-like crossover and mutation. Further variables were population size, mutation rate and the crossover strategy. In order to eliminate the stochastic differences between different GA runs, each parameter set was used in 100 parallel runs and the averages and standard deviations for the various results were calculated and displayed on the following figures. All results were compared with a "random" screening method that was simulated by a repeated random selection of new products from the library.

Crossover between bit strings that encode the starting materials A, B and C, or crossover that allows cutting within these bit strings, can be compared when using binary bit strings. While the first strategy of replacing only whole building blocks appeals more to chemists, the latter is more similar to crossover in DNA. Population sizes of 5, 10, 20 or 80 and stochastic mutation rates of 1 or 10% were investigated in this case (figure 7). The GA-driven selection of new members was able to find considerably more active compounds within the first few generations than "random" screening. The slopes of the GA curves first increase due to the "learning" process and decrease again when many of the limited number of active compounds have been discovered.

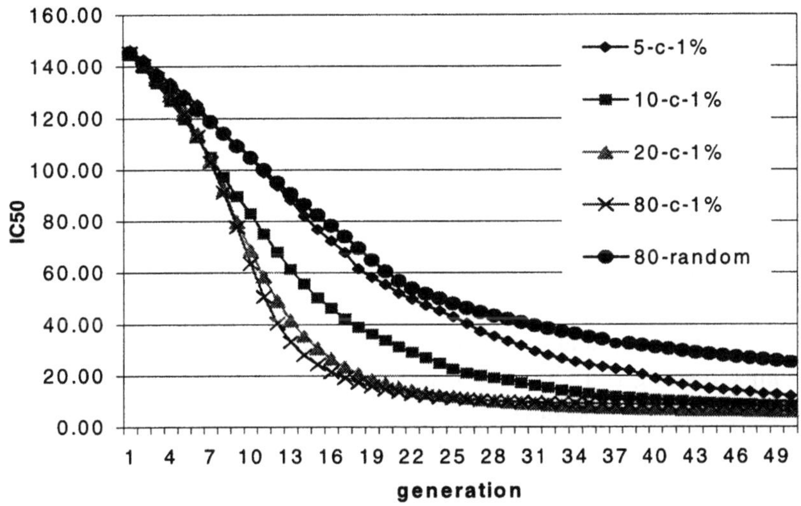

Figure 7. The influence of the generation size **n** on the average fitness of the **n** best parents at a given generation. **n** was set to 5, 10, 20 or 80 using a binary genome and crossover by cutting between starting material genes (option c) and a fixed stochastic mutation rate of 1%. Random selection of 80 new products as opposed to the GA driven selection is shown by the curve 80-random.

To quantify the benefit of an evolutionary GA strategy, as opposed to random screening, we introduced a performance criterion

$$P = (m_{GA} / n_{GA})/(m_{random} / n_{random}) \quad (1)$$

where m_{GA} and m_{random} are the slopes of the performance curves for the GA and random screening normalised to the respective population sizes. P represents the average activity increase that was achieved by an individual genome in one evolutionary cycle using the given GA parameters. The maximum value P_{max} for generation sizes of 5, 10, 20 and 80 was 23.6, 14.1, 10.3, and 3.1, respectively. In other words, small populations "learn more" per individual, whereas larger populations "learn faster" as shown by P_{max}*n that is 118, 141, 206 and 248, respectively.

Cutting the bit strings in a chemically meaningful way only between the building blocks A, B and C (option c), showed a slightly better performance as opposed to allowing for a DNA-like crossover at any bit. The latter strategy will result in children with higher diversity since starting materials may be used that were not part of the parent products (see section 3.1.1). Similarly, increasing the mutation rate from 0.1 to 10% lowers the performance of the GA by "destroying" the acquired knowledge about the fitter bit strings. The best performing GA parameter set for binary encoding used the lowest meaningful 0.1% stochastic mutation rate and cutting only between building blocks (strategy 20-c-1%, figure 8).

Another parameter of evolutionary optimisation is the time taken until the best solutions are found. Figure 8 displays the activity of the most active product found during the course of a given GA by using various parameter sets and averaged over 100 parallel runs each. Thus, strategies 20-c-1% and 80-c-1% discover products with an activity below 1 μM with a probability of 90% in generation 30 and 13, respectively. This corresponds to the synthesis of 600 compounds for a generation size 20 and 1040 for **n**= 80. The average number of genomes that are needed to find one of these 9 best products by random screening is 15360/9 = 1707; therefore, the performance of the GA in finding the best products is 284% (1707/600) and 164% (1707/1040) higher as compared to random screening. Again, the performance of the GA is better per individual when using smaller generation sizes and lower when using a high mutation rate.

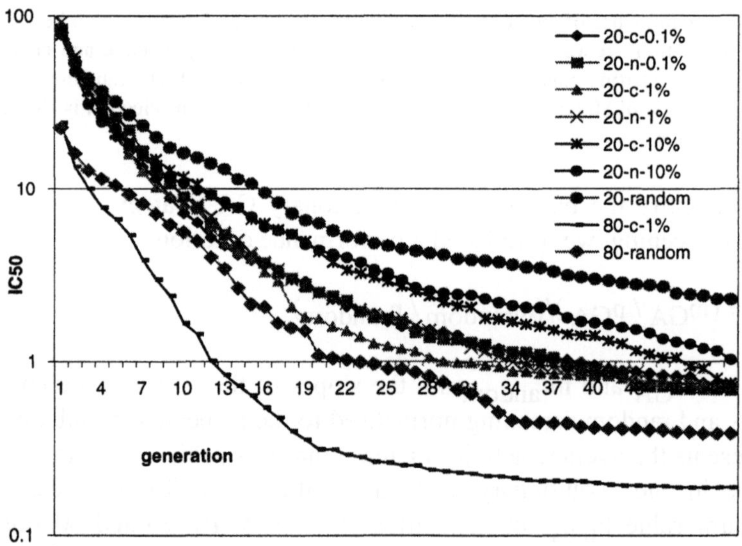

Figure 8. The activity of best reaction product found by the GA during the course of evolution depending on the stochastic mutation rate, crossover and generation size. The results are displayed as averages from 100 parallel runs for each GA parameter set and compared to random selection (20- and 80-random).

Genetic algorithms learn in an implicit, heuristic manner whereby learning means that a new generation should be better then the previous one. Figure 9 displays the average activity of the selected **n** new genomes for the best c-1% strategy. The average activity of randomly selected new compounds (80-random) remains the same since no learning takes place. The average activity for the GA driven selection, however, increases at the beginning of the evolution process - a measure for the implicit learning process during a given GA.

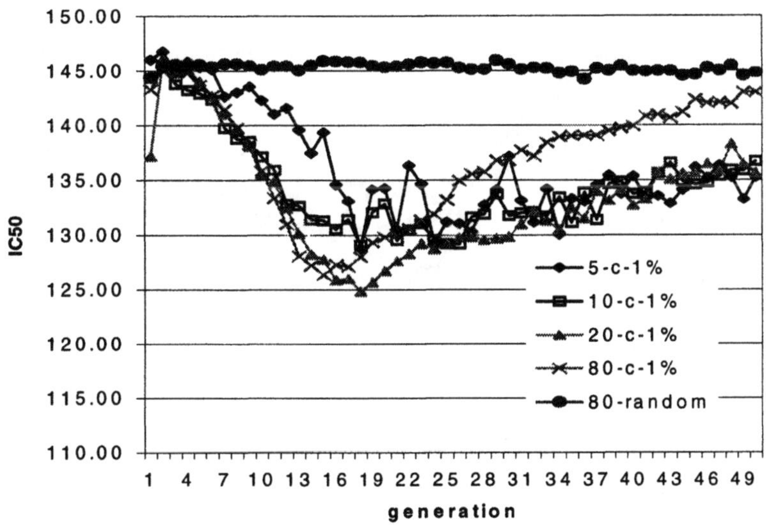

Figure 9. The average activity of the selected N new children at each new generation using 1% mutation rate and crossover between adducts.

After reaching a maximum slope the learning decreases, due to the increasing depletion of active products in our search space. While the shapes of the respective curves are rather similar, the slope and generation number of the average activity maximum provide a tool to assess the learning process during a GA driven evolution. Thus, the 20-c-1% strategy seems to be optimal both with respect to the number of used individuals, the learning slope and the achieved maximum activity at generation 18.

This learning behaviour was also studied using decimal encoding of building blocks in the genome of the Ugi three component products. As depicted in figure 10, a mutation rate of 100% corresponds to random screening - no learning takes place (series D20-100%). It is further interesting to note that starting with a 5% stochastic mutation rate up to 20% does not yield a significantly decreased learning rate. This behaviour illustrates that mutation takes on the role of supplying new possibilities for optimisation whereas learning is a pure effect of crossover. Overall, the binary encoding strategy (B20-c-1%) seemed to be a little more effective with respect to learning.

Since the DNA-like crossover with a binary bit string is a mixture of mutation and crossover, the mutation rate for the decimal encoding strategy has a different meaning and effect when compared to mutation in binary bit strings.

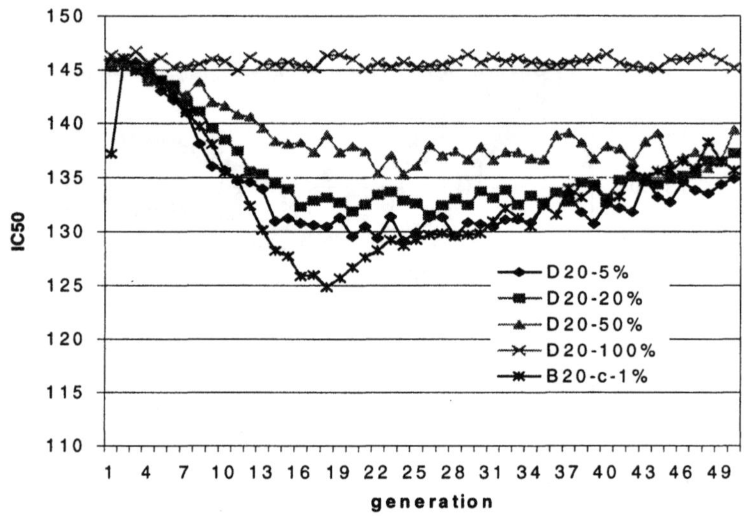

Figure 10. The average activity of the selected n new children at each new generation using various mutation rates and decimal encoding (series D) as compared to binary encoding (series B) with crossover between building blocks.

Intrinsically, the mutation rate has to be set to a higher value when decimal encoding is used instead of binary encoding. The advantage of decimal encoding is that one can clearly separate the effects of crossover and mutation. In the current implementation, the given mutation probability of product **A-B-C** was equally distributed between **A**, **B** and **C**, irrespective of how many possible building blocks in each list there were. Mutation can act either stochastically, or by making the chance of changing any **A**(x) to any **A**(y) the same. In the nearest neighbours method, the probability of arriving at **A**(x+1) from starting with **A**(x) is set to a higher value.

A connected important issue for genetic algorithms might be the order of the encoded building blocks. In the previous examples, the order for all building blocks was completely accidental as shown for **B** in figure 6. It may be helpful to pre-sort **A**, **B** and **C** on the axis of the search space according to their structural similarity. Furthermore, nearest neighbour's mutation should have a beneficial effect, assuming that not only chemically similar molecules are clustered in the 4-dimensional search space but also the biological activities. This clustering should allow for faster optimisation.

To prove or disprove this theory, we have ordered all starting material according to their Tanimoto similarity using the Daylight software [24]. From the viewpoint of a chemist the obtained order **B1** to **B10** indeed seems to make sense as shown in figure 11 and opposed to the order in figure 6.

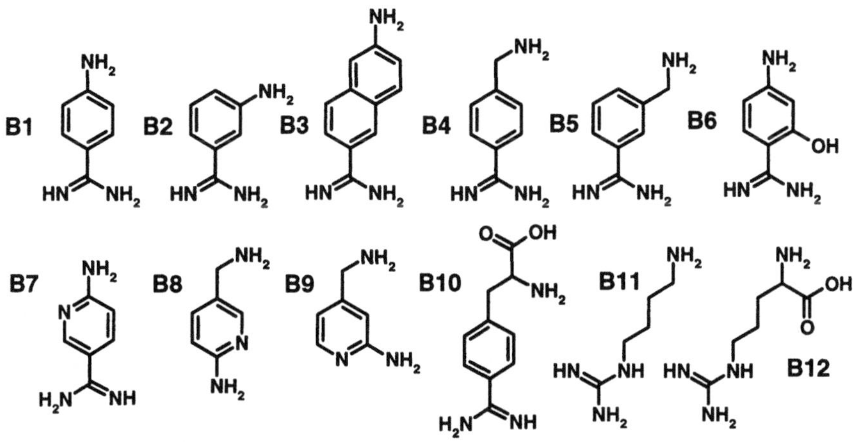

Figure 11. Amines B that were resorted according to the Tanimoto similarity and used for running the GA on a resorted structure-activity landscape.

It has to be noted that for the binary encoding strategy with a larger number of building blocks sorting is not meaningful, since mutations in different positions of the particular building block bit string will yield different results. On the DNA that uses only four building blocks, the nearest neighbour strategy is partly possible and seems to work. Thus, a single mutation of UUC (phenylalanine) to UUA (leucine) yields a very similar building block with respect to hydrophobicity, whereas 3 mutations are required for CAA that encodes glutamate, a very different amino acid with respect to both leucine and phenylalanine. The statistics of this encoding provides a metric that has been used by Schneider [25] to predict proteins with desired biological activities using evolutionary algorithms and neuronal networks.

The differences between random screening and the stochastic mutation using binary (B20-c-1%, with cutting between building blocks only, 20% mutation rate) or decimal encoding (D20-20% with 20% mutation rate) strategy were compared with the decimal encoding + nearest neighbour mutation on the unsorted (D20-20%N) and resorted (D20-20%re-sorted_N) structure activity landscapes. Surprisingly, there was almost no difference between regarding the most active compounds during the GA (figure 12)!

The D20-20% strategy performs as well on the unsorted as well as on the re-sorted landscape with, or without, nearest neighbour mutations. This observation seems to point out that a structural similarity based on simple building block pre-sorting has no, or at least a negligible effect although the biological activities do cluster to a certain extent on the sorted landscape in our example.

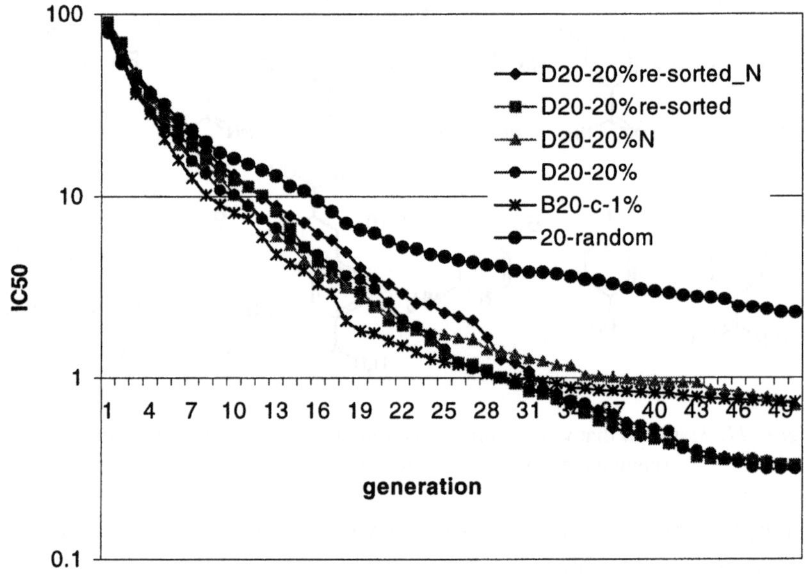

Figure 12. The average activity of the best **n** new children at each new generation using decimal encoding (series D) and binary encoding (series B) and comparing the effect of building block resorting (re-sorted) according to building block similarity. Strategy D20-20%re-sorted_N refers to a population size of **n**=20, decimal encoding, 20% mutation rate with nearest neighbours selection and building blocks similarity sorting.

This somewhat astonishing behaviour was verified further using other data sets of similar size and seems to depend on the structure of these multidimensional search spaces of building blocks and biological data. The reason may be found in the fractal nature of such large databases. This fractal self-similarity does not change to a significant degree upon resorting.

4. DIVERSITY IN LARGE LIBRARIES

Evolution in nature generates diversity by changing an essential amino acid required for biological activity by a single mutation on the DNA, mutation at a less important site yields a similar protein. By analogy, genetic algorithms may be used to generate simultaneous diversity and similarity by mutation and crossover with respect to the applied selection function.

An optimally diverse compound library for biological random screening would be a collection of representatives from a variety of compound clusters of similar molecules. Thus, the first step is to cluster molecules according to their similarity and composing afterwards a diverse set. The central task to solve this problem, for all kinds of molecules, is the identification of suitable molecular descriptors.

Using a GA for generation of diversity, one may chose a molecular property and select for molecules that are different with respect to that property. A dissimilar, unique molecular weight is such a property that would facilitate the deconvolution of combinatorial library mixtures by mass spectrometry. The optimal design of such mixtures was the target function in a recent application of a GA [26].

The diversity of combinatorial libraries appears to be better accessible by computational methods as opposed to that of general molecules due to their well defined "closed" chemical space. Thus, several methods have been introduced to design diverse combinatorial compound libraries by selecting optimal building blocks for synthesis [27,28]. Using molecular volume, lipophilicity, charge and H-bond donor or acceptor descriptors [29], it was shown that peptoid-type combinatorial libraries may be designed to exhibit the same density of these structural fragments counted per atom [30] as commercial drugs. Other new methods include the application of information theory and stochastic algorithms to aid selecting optimally diverse compounds [31] for biological screening. The implicit, but unproved, assumption is that one may find therefore all desired biological activities within this library.

REFERENCES

1. Cook, W.J., Cunningham, W.H., Pulleyblank, W.R. and Schrijver, A. *Combinatorial Optimization*, Wiley, 1997.
2. Ugi, I., Bauer, J., Bley, K., Dengler, A., Dietz, A., Fontain, E., Gruber, B., Herges, R., Knauer, M., Reitsam, K. and Stein, N. Computer-assisted solution of chemical problems - the historical development and the present state of the art of a new discipline of chemistry. *Angew. Chem. Int. Ed. Engl.*, 1993, **32**, 201-227.

3. Ugi, I., Wochner, M., Fontain, E., Bauer, J., Gruber, B. and Karl, R. Chemical similarity, chemical distance and computer assisted formalized reasoning by analogy. In *Concepts and Applications of Molecular Similarity*, Eds Johnson, M.A., Maggiora, G.M., John Wiley & Sons Inc, New York, 1990, pp. 239-288.
4. Weininger, D. SMILES: a chemical language and information system. 1. Introduction to methodology and encoding rules. *J. Chem. Inf. Comput. Sci.,* 1988, **28**, 31-36.
5. Venkatasubramanian, V., Chan, K. and Caruthers, J. Evolutionary Design of molecules with desired properties using the genetic algorithm. *J. Chem. Inf. Comput. Sci.,* 1995, **35**, 188-195.
6. Glen, R.C. and Payne, A.W.R. A genetic algorithm for the automated generation of molecules within constraints. *J. Comput.-Aid. Mol. Des.,* 1995, **9**, 181-202.
7. Weber, L., Wallbaum, S., Broger, C., and Gubernator, K. A genetic algorithm optimizing biological activity of combinatorial compound libraries. *Angew. Chem. Int. Ed. Engl.,* 1995, **107**, 2453-2454.
8. Wild, D.J. and Willet, P. Similarity Searching in Files of Three-Dimensional Chemical Structures. Alignment of Molecular Electrostatic Potential Fields with a Genetic Algorithm. *J. Chem. Inf. Comput Sci.,* 1996, **36**, 159-167.
9. Sheridan, R.P. and Kearsley, S.K. Using a genetic algorithm to suggest combinatorial libraries. *J. Chem. Inf. Comput. Sci.,* 1995, **35**, 310-320.
10. Rechenberg, I. *Evolutionsstrategie: Optimierung technischer Systeme nach Prinzipien der biologischen Evolution,* Frommann-Holzboog, Stuttgart, 1973.
11. Holland, J.H. *Adaptation in Natural and Artificial Systems,* The University of Michigan Press, Ann Arbor, MI, 1975.
12. Clark, D.E. Some Current Trends in Evolutionary Algorithm in Research Exemplified by Applications in Computer-Aided Molecular Design. *Communications in Mathematical and in Computer Chemistry (MATCH)*, 1998, **38**, 85-93.
13. Weber, L. Evolutionary combinatorial chemistry: application of genetic algorithms, *Drug Discovery Today*, 1998, **3**, 379-385.
14. Desmond, J.M. Applications of soft computing in drug design. *Exp. Opin. Ther. Patents,* 1998, **8**, 249-258.
15. Goldberg, D.E. *Genetic Algorithms in Search, Optimization and Machine Learning,* Addison-Wesley, Reading, MA, 1989.
16. Baeck, T., Fogel, D.B. and Michalewicz, Z. *Handbook of Evolutionary Computation,* IOP Publishing and Oxford University Press, Bristol/New York, 1997.
17. Holland, J.H. *Hidden Order - How Adaptation Builds Complexity,* Addison-Wesley, Reading, MA, 1996.
18. Forrest, M. and Mitchell, M. Relative building-block fitness and the building-block hypothesis. In *Foundations of Genetic Algorithms 2*, Ed. Whitley, D., Morgan Kaufmann, San Mateo, CA, 1993, pp. 109-126.
19. Gobbi, A. and Poppinger, D. Genetic Optimization of Combinatorial Libraries. *Biotechnol. Bioeng.,* 1998, **61**, 47-54.
20. Reetz, M.T., Zonta, A., Schimossek, K., Liebeton, K. and Jaeger, K.-E. Creation of Enantioselective Biocatalysts for Organic Chemistry by In Vitro Evolution. *Angew. Chem.,* 1997, **109**, 2961-2963.
21. Yokobayashi, Y., Ikebukuro, K., McNiven, S. and Karube, I. Directed evolution of trypsin inhibiting peptides using a genetic algorithm. *J. Chem. Soc., Perkin Trans. I,* 1996, 2435-2437.
22. Singh, J., Ator, M.A., Jaeger, E.P., Allen, M.P., Whipple, D.A., Soloweij, J.E., Chowdhary, S. and Treasurywala, A.M. Application of genetic algorithms to

combinatorial synthesis: a computational approach to lead identification and lead optimization. *J. Am. Chem. Soc.,* 1996, **118**, 1669-1676.

23. Illgen, K., Enderle, T., Clemens Broger, C. and Weber, L. Simulated Molecular Evolution in a Full Combinatorial Library. *Chemistry & Biology,* 1998, in press.
24. James, C.A. and Weininger, D. *Daylight Theory Manual.* Daylight Chemical Information Systems Inc., Irvine, 1995.
25. Schneider, G., Grunert, H.P., Schuchhardt, J., Wolf, K.-U., Müller, G., Habermehl, K.-O., Zeichhardt, H. and Wrede, P. A peptide selection scheme for systematic evolutionary design and construction of synthetic peptide libraries. *Minimal Invasive Medizin,* 1995, **6**, 106-115.
26. Brown, R.D. and Martin, Y.C. Designing combinatorial library mixtures using a genetic algorithm. *J. Med. Chem.*, 1997, **40**, 2304-2313.
27. Good, A.C. and Lewis, R.A. New Methodology for Profiling Combinatorial Libraries and Screening Sets: Cleaning Up the Design Process with HARPick. *J. Med. Chem.,* 1997, **40**, 3926-3936.
28. Gillet, V.J., Willett, P. and Bradshaw, J. The Effectiveness of Reactant Pools for Generating Structurally-Diverse Combinatorial Libraries. *J. Chem. Inf. Comput. Sci.,* 1997, **37**, 731-740.
29. Shemetulskis, N.A., Dunbar, J.B. Jr., Dunbar, B.W., Moreland, D.W. and Humblet, C. Enhancing the diversity of a corporate database using chemical clustering and analysis. *J. Comput.-Aided Mol. Des.*, 1995, **9**, 407-416.
30. Martin, E.J., Blaney, J.M., Siani, M.A., Spellmeyer, D.C., Wong, A.K. and Moos, W.H. Measuring diversity: experimental design of combinatorial libraries for drug discovery. *J. Med. Chem.,* 1995, **38**, 1431-1436.
31. Agrafiotis, D.K. Stochastic Algorithms for Maximizing Molecular Diversity. *J. Chem. Inf. Comput. Sci.,* 1997, **37**, 841-851.

Chapter 6

Subset-Selection Methods For Chemical Databases

Methods for Subset Selection

P. WILLETT
Department of Information Studies, University of Sheffield, Western Bank, Sheffield S10 2TN, UK. Email: p.willett@sheffield.ac.uk

Key words: Clustering, Dissimilarity, Partitioning, Optimisation, Similarity, Validation Methods.

Abstract: This chapter reviews the methods that are available for selecting subsets of chemical databases, both real and virtual. Cluster-based, dissimilarity-based, partition-based and optimisation-based algorithms are discussed are compared in terms of their efficiency and effectiveness, and of their applicability to a range of subset-selection tasks.

1. INTRODUCTION

Recent developments in combinatorial chemistry and high-throughput screening (HTS) have revolutionised the methods that are used for lead discovery in the pharmaceutical and agrochemical industries. As this book makes clear, the successful application of these technological developments has necessitated the introduction of novel computer techniques for the analysis and processing of the large amounts of structural and biological data that result from lead-discovery programmes. In this chapter we review those techniques that have been developed for the selection of structurally diverse subsets of files of molecules, although we note in passing that many of these techniques are also applicable to related tasks such as the identification of structural overlap in databases and the mapping of structural space.

The new subset-selection methods draw heavily upon the techniques that are used for searching and clustering databases of two-dimensional (2D) and three-dimensional (3D) chemical structures [1-4]. These techniques were developed for processing databases containing individual compounds that have been reported in the literature, that have been synthesised in-house within an organisation, or that have been made available by a commercial compound supplier. However, the techniques are equally applicable to *virtual databases* obtained by enumerating the products of a combinatorial synthesis. In the following, unless stated otherwise, we shall not be concerned with the precise nature of the database that is being processed, focusing the discussion on the methods that are now being used to select diverse database subsets: note the emphasis on *methods*, since there may often be several different *algorithms* for the implementation of the same method (as with the stored-matrix and RNN algorithms for certain classes of hierarchic clustering method, as discussed in the next section of this chapter).

The implementation of a selection method may require the specification of the *molecular descriptors* and the *inter-molecular similarity measures* that are to be used. Firstly, the structures in the database must be characterised by some type of descriptor that can be rapidly generated from a machine-readable structure representation, this typically being a 2D connection table or a set of experimental or calculated 3D co-ordinates. There is much current interest in the development and comparison of descriptors for diversity analysis (see, *e.g.*, [5-9]): here, we note merely that the two most important types of descriptor that have been used thus far are *fingerprints* (bit-strings encoding the presence of fragment substructures within a molecule) and *physical properties*. Secondly, given an appropriate descriptor, some means must be found to quantify the degree of *similarity*, *dissimilarity* or *distance* between pairs (or larger groups) of molecules [10, 11]. This is normally done using the measures of inter-molecular structural similarity that have been developed for similarity searching in chemical databases [2].

The choice of representation, of similarity measure and of selection method are not independent of each other. For example, some types of similarity measure (specifically the association coefficients as exemplified by the well-known Tanimoto coefficient) seem better suited than others (such as Euclidean distance) to the processing of fingerprint data [12]. Again, the partition-based methods for compound selection that are discussed below can only be used with low-dimensionality representations, thus precluding the use of fingerprint representations (unless some drastic form of dimensionality reduction is performed, as advocated by Agrafiotis [13]). Thus, while this chapter focuses upon selection methods, the reader should keep in mind the representations and the similarity measures that are being used: recent, extended reviews of these two important components of diversity analysis are provided by Brown [14] and by Willett *et al.* [15].

There are many different types of selection operation that one might wish to carry out, including: the selection of a small number of reagents from those listed in a commercial catalogue such as the *Available Chemicals Directory* (ACD) [16] for a combinatorial synthesis; the selection of individual compounds from an existing database for testing in a high-throughput screen; and the selection for synthesis and testing of a combinatorial subset from a fully enumerated virtual library. In what follows we shall assume that the selection methods described can be applied to any of these problems, unless stated otherwise.

There is already an extensive literature relating to compound-selection methods, from which it is possible to identify four major classes of method although, as we shall see, there is some degree of overlap between these four classes, *viz cluster-based* methods, *dissimilarity-based* methods, *partition-based* methods and *optimisation* methods. The next four sections of this chapter present the various algorithms that have been suggested for each approach; we then discuss comparisons and applications of these algorithms, and the chapter concludes with some thoughts on further developments in the field.

2. CLUSTER-BASED SELECTION METHODS

Cluster analysis, or *clustering*, is the process of subdividing a group of objects (chemical molecules in the present context) into groups, or *clusters*, of objects that exhibit a high degree of both intra-cluster similarity and inter-cluster dissimilarity [17, 18]. It is thus possible to obtain an overview of the range of structural types present within a dataset by selecting one, or some small number, of the molecules from each of the clusters resulting from the application of an appropriate clustering method to that dataset. The representative molecule (or molecules) for each cluster is either selected at random or selected as being the closest to the centre of that cluster.

Very many different clustering methods have been described in the literature, and a considerable amount of effort has gone into comparing the effectiveness of the various methods for clustering chemical structures (see, *e.g.*, [6, 19-21]). Clustering methods can produce *overlapping* clusters, in which each object may be in more than one cluster, or *non-overlapping* clusters, in which each object occurs in only one cluster. Of these, the latter are far more widely used and thus most of the methods that are discussed below belong to this class; an example of an overlapping method that has been used for compound selection is provided by the work of the group at the National Cancer Institute [22, 23]. There are two main classes of non-overlapping cluster methods: *hierarchical* methods and *non-hierarchical* methods.

An hierarchical clustering method produces a classification in which small clusters of very similar molecules are nested within larger and larger clusters of less closely-related molecules. Hierarchical *agglomerative* methods generate a classification in a bottom-up manner, by a series of agglomerations in which small clusters, initially containing individual molecules, are fused together to form progressively larger clusters. Conversely, hierarchical *divisive* methods generate a classification in a top-down manner, by progressively sub-dividing the single cluster which represents the entire dataset [17, 18]. The agglomerative methods are far more widely used and comparative experiments [6, 20] have demonstrated their effectiveness for clustering chemical structures; they are hence discussed further below.

The hierarchic agglomerative methods can all be implemented by means of the basic algorithm shown in Figure 1, where a *point* is either a single molecule or a cluster of molecules. This procedure is known as the *stored matrix* algorithm since it involves random access to the inter-molecular similarity matrix throughout the entire cluster-generation process. Individual hierarchical agglomerative methods differ in the ways in which the most similar pair of points is defined and in which the merged pair is represented as a single point. A general formulation for the stored-matrix algorithm that encompasses all the common methods (such as the *single linkage*, *complete linkage* and *centroid* methods) is described by Lance and Williams [24]. Murtagh [25] discusses the *reducibility property*. If a method satisfies this property then agglomerations can be done in restricted areas of the similarity space and the results amalgamated to form the overall hierarchy; moreover, in such cases, the stored-matrix algorithm can be replaced by the computationally more efficient *reciprocal nearest neighbour* (RNN) algorithm. In this, a path is traced through the similarity space until a pair of points is reached that are more similar to each other than they are to any other points, *i.e.*, they are RNNs. These RNN points are fused to form a single new point, and the search continues until the last unfused point is reached. The basic RNN algorithm is thus as shown in Figure 2, where NN(X) denotes the nearest neighbour for the point X, and the final, overall hierarchy is then created from the list of RNN fusions that has taken place. This approach is applicable to clustering methods in which the most similar pair at each stage is defined by a distance measure, as in the popular method first described by Ward [26]. Detailed discussions of the applicability of the RNN algorithm are provided by Brown and Martin [6] and by Downs and Willett [21].

1. Calculate the inter-molecular similarity matrix.

2. Find the most similar pair of points in the matrix and merge them into a cluster to form a new single point.

3. Calculate the similarity between the new point and all remaining points.

4. Repeat Steps 2 and 3 until only a single point remains, *i.e.*, until all of the molecules have been merged into one cluster.

Figure 1. Stored matrix algorithm for hierarchic agglomerative clustering methods.

1. Mark all molecules, *I*, as unfused.

2. Starting at an unfused *I*, trace a path of unfused nearest neighbours (NN) until a pair of RNNs is encountered, *i.e.*, trace a path of the form J := NN(I), K := NN(J), L := NN(K)..... until a pair is reached for which Q = NN(P) and P = NN(Q).

3. Add the RNNs *P* and *Q* to the list of RNNs along with the distance between them, mark *Q* as fused and replace the centroid of *P* with the combined centroid of *P* and *Q*.

4. Continue the NN-chain from the point in the path prior to *P*, or choose another unfused starting point if *P* was a starting point.

5. Repeat Steps 2-4 until only one unfused point remains.

Figure 2. Reciprocal nearest neighbours algorithm for hierarchic agglomerative clustering methods that satisfy the reducibility property.

Once the cluster hierarchy has been produced, some means is required to identify a set of clusters from which molecules can be selected. This is normally achieved by applying a threshold similarity to the hierarchy and identifying the clusters present in the resulting partition (*i.e.*, a set of non-overlapping groups having no hierarchical relationships between them) of the dataset. A non-hierarchical method, conversely, generates a partition of a dataset directly. There are a combinatorial number of possible partitions, making a systematic evaluation of them totally infeasible, and many different heuristics have thus been described to allow the identification of good, but possibly sub-optimal, partitions. The three most common types of non-hierarchic method are the *single-pass*, *relocation* and *nearest-neighbour* methods, all of which are less demanding of computational resources than the hierarchical methods.

Single-pass methods are simple to implement and very fast. As the name suggests, they require a single pass through the dataset to assign compounds to clusters, with a threshold of similarity being used to decide whether to assign the next compound to an existing cluster or to use it to commence a new cluster. A large-scale application of the single-pass method has been described by Hodes and co-workers [22, 23] but the inherent order-dependent nature of the processing means that it is not widely used [21]. Relocation methods assign compounds to a user-defined number of *seed* clusters and then iteratively reassign compounds to see if better (in some sense) clusters result [27]. These methods, such as the *k-means* method [18, 28, 29], have been used for chemical applications but are prone to reaching local optima rather than the global optimum, and it is generally not possible to determine when, or whether, this optimal solution has been reached. Instead, the most widely used non-hierarchical method, indeed probably the single most widely used method, for cluster-based compound selection is the near neighbours *Jarvis-Patrick* method [30].

1. Identify the top-K nearest neighbours for each of the N molecules in the dataset.

2. Create an N-element array, *Label*, that contains a cluster label for each of the N molecules in the dataset. Initialise *Label* by setting each element to its array position, thus assigning each molecule to its own initial cluster;

3. For each pair of molecules, I and J ($I < J$),

   If the pair have at least K_{min} of their top-K nearest neighbours in common and each is in the top-K nearest-neighbour list of the other, then replace all occurrences of the *Label* entry for J with the *Label* entry for I.

4. The members of each cluster then all have the same entry in the final *Label*.

Figure 3. Algorithm for the Jarvis-Patrick clustering method

The Jarvis-Patrick method involves the use of a list of the top K *nearest neighbours* for each molecule in a dataset, *i.e.*, the K molecules that are most similar to it. Once these lists have been produced for each molecule in the dataset that is to be processed, two molecules are clustered together if they are nearest neighbours of each other and if they additionally have some minimal number of nearest neighbours, K_{min}, in common. The algorithm is detailed in Figure 3. With the relocation methods, the user has to specify the number of initial cluster seeds: here, the partition that is obtained is governed by the choice of K_{min}, and it is hence generally necessary to experiment with a range of K_{min} values until roughly the required number of clusters is obtained. However, even this may not suffice since the method can often result in a small number of clusters containing very many molecules and/or a large number of *singletons*, *i.e.*, clusters that contain just a single molecule. Menard *et al.* [31] have recently described a variant, which they refer to as *cascaded clustering*, in which the method is applied repeatedly so as to produce a well-balanced pseudo-hierarchy that avoids these problems. Other variations include building the nearest neighbour table using a similarity cut-off, rather than a fixed number of nearest neighbours, and not requiring that two molecules need to be in each other's lists if they are to be clustered together. Many examples of the use of the Jarvis-Patrick method for compound selection, coupled in some cases with detailed discussions of the method's strengths and weaknesses, have been reported in the literature [31-37].

3. DISSIMILARITY-BASED SELECTION METHODS

Dissimilarity-based methods seek to identify a subset comprising the n most diverse molecules in a dataset containing N molecules (where, typically, $n << N$). There are no less than w size-n subsets that can be generated from a size-N dataset, where

$$w = \frac{N!}{n!(N-n)!} \tag{1}$$

and the identification of the optimally diverse one is thus computationally infeasible, unless both n and N are very small. Practicable approaches hence involve approximate methods that are not guaranteed to result in the identification of the most dissimilar possible subset [38-45]; that said, there is evidence to suggest that the subsets identified are only marginally sub-optimal [46] when the diversity is quantified by means of a diversity index, such as those described by Gillet [47]. Thus far, two major classes of algorithm have been described: *maximum-dissimilarity* algorithms and *sphere-exclusion* algorithms [48].

The basic maximum-dissimilarity algorithm for selecting a size-*n* *Subset* from a size-*N* *Dataset* is shown in Figure 4. This algorithm, which was first described by Kennard and Stone [49] almost three decades ago and which was applied to compound selection by Lajiness and Bawden (see, *e.g.*, [50, 51]), permits many variants depending upon the precise implementation of Steps 1 and 3. Possible mechanisms for the choice of the initial compound in Step 1 include: choosing a compound at random; choosing that compound that is most dissimilar to the other compounds in *Dataset*; or choosing that compound that is nearest to the centre (in some sense) of *Dataset*, *inter alia*. Step 3 in the figure requires a quantitative definition of the dissimilarity between a single compound in *Dataset* and the group of compounds that comprise *Subset*, so that the most dissimilar molecule can be identified in each iteration of the algorithm. There are several ways in which "most dissimilar" can be defined, with each definition resulting in a different version of the algorithm and hence in the selection of a different subset [52] (in just the same way as different hierarchic agglomerative clustering methods result from the use of different similarity criteria in the Lance-Williams formula for the stored-matrix algorithm [24]). Efficient examples of such definitions that have been described in the literature for dissimilarity-based selection include *MaxSum* [40, 53] and *MaxMin* [29, 41]. Let *DIS*(*A*,*B*) be the dissimilarity between two molecules, or sets of molecules, *A* and *B*. Consider a single compound, *J*, taken from *Dataset* and the *m* compounds that form the current membership of *Subset* at some stage in the selection process; then the dissimilarity between *J* and *Subset*, *DIS*(*J*, *Subset*), is given by

$$\sum DIS(J,K) \text{ and } minimum\{DIS(J,K)\} \qquad (2)$$

in the case of the MaxSum and MaxMin definitions, respectively, with K ($1 \leq K \leq m$) ranging over all of the m molecules in *Subset* at that point. The molecule chosen for addition to *Subset* is then that with the largest value of *DIS*(*J*, *Subset*).

1. Initialise *Subset* by transferring a compound from *Dataset*.
2. Calculate the dissimilarity between each remaining compound in *Dataset* and the compounds in *Subset*.
3. Transfer to *Subset* that compound from *Dataset* that is most dissimilar to *Subset*.
4. Return to Step 2 if there are less than *n* compounds in *Subset*.

Figure 4. General maximum-dissimilarity algorithm

A further variant of the basic approach is to specify a threshold dissimilarity, *t*, and then to reject the molecule selected in Step 2 if it has a dissimilarity less than *t* with any of the compounds already in *Subset*. The inclusion of such a threshold results in a maximum dissimilarity algorithm that is not too far removed from the basic sphere-exclusion approach [41, 54]. Here, a threshold *t* is set, which can be thought of as the radius of a hypersphere in multi-dimensional chemistry space. A compound is selected, either at random or using some rational basis, for inclusion in *Subset* and the algorithm then excludes from further consideration all those other compounds within the sphere centred on that selected compound, as shown in Figure 5. Many variants are again possible, depending upon the manner in which Stage 2 is implemented. Thus, one can choose that molecule that is most dissimilar to the existing *Subset*, in which case different results will be obtained (as with the maximum dissimilarity algorithms) depending upon the dissimilarity definition that is adopted. Alternatively, a compound can be selected at random, as in the MDISS and DIVPIK programs [43, 44]; this results in an exceptionally fast, but non-deterministic, algorithm.

1. Define a threshold dissimilarity, *t*.

2. Transfer a compound, *J*, from *Dataset* to *Subset*.

3. Remove from *Dataset* all compounds that have a dissimilarity with *J* of less than *t*.

4. Return to Step 2 if there are compounds remaining in *Dataset*.

Figure 5. General sphere-exclusion algorithm

The close relationship that exists between these two classes of algorithm has recently been highlighted by Clark [45], who describes a program called OptiSim (for Optimizable K-Dissimilarity Selection) that is summarised in Figure 6 and that makes use of an intermediate pool of selected compounds, here called *Subsample*. An inspection of this figure shows that the mode of processing is determined by the value of *K* (the size of *Subsample*) that is specified, with values of *K* equal to 1 and to *N* corresponding to (versions of) sphere-exclusion and maximum dissimilarity, respectively. Clark presents a detailed discussion of how the choice of *K* affects the behaviour of the algorithm and the trade-offs that are to be expected between what he describes as the *representativeness* of subsets generated by sphere-exclusion methods and the *diversity* of subsets generated by maximum-dissimilarity methods.

1. Define a threshold dissimilarity, *t*.

2. Initialise *Subset* by transferring a compound from *Dataset*.

3. Select a compound, *J*, from *Dataset*. If it has a dissimilarity less than *t* with any compound in *Subset* then remove it from *Dataset*; otherwise add it to *Subsample*.

4. Repeat Step 3 until *Subsample* contains *K* molecules.

5. Transfer to *Subset* that compound from *Subsample* that is most dissimilar to *Subset*. Return the remaining members of *Subsample* to *Dataset*.

6. Return to Step 3 if there are less than *n* compounds in *Subset*.

Figure 6. OptiSim algorithm [44]

The use of a dissimilarity threshold ensures that no two molecules in *Subset* will be strongly similar to each other. Gardiner *et al.* [55] have recently described a novel algorithm that allows the identification of all subsets satisfying this dissimilarity criterion, rather than the single subset that is the result of the other algorithms discussed thus far. The algorithm is summarised in Figure 7 and derives from the observation that subset detection is equivalent to the NP-complete problem of identifying all of the *cliques* in a graph, where a clique is a subgraph in which every vertex is connected to every other vertex and which is not contained in any larger subgraph with this property. Each node in a *subset-selection graph* denotes one of the molecules in the dataset from which the subsets are to be selected and each edge, *IJ*, is set to 0 or 1 depending on whether *I* and *J* have a dissimilarity greater than or less than *t*. Gardiner *et al.* report a comparison of several different clique-detection algorithms for processing subset-selection graphs, and suggest that one due to Babel [56] is sufficiently fast to enable the procedure to be applied to the selection of reagents for combinatorial synthesis, *e.g.*, selecting 40 diverse amines from amongst those available in the ACD. Once all of the subsets have been generated by the procedure shown in Figure 7, a further filtering step, based on criteria such as cost, physicochemical-parameter or diversity-index values, is needed to identify the particular subset that will be chosen for use in some application. The satisfaction of a quantitative criterion also underlies some of the optimisation-based approaches to compound selection that are discussed later in this chapter.

1. Define a threshold dissimilarity, *t*.

2. Generate an $N \times N$ dissimilarity matrix in which $M(I,J)$ contains the dissimilarity between molecules *I* and *J*.

3. Generate a graph, *G*, from *M* by setting each element $M(I,J)$ to one (or zero) if it is greater than (or not greater than) *t*.

4. Use a clique-detection algorithm to identify the set of size-*n* cliques in *G*.

Figure 7. Clique-based processing to identify all subsets meeting a dissimilarity criterion

4. PARTITION-BASED SELECTION METHODS

Partition-based selection requires the identification of a set of *P* characteristics that span the chemical space of interest [57], these characteristics normally being molecular properties that would be expected to affect binding at a receptor site. The range of values for each such characteristic, I ($1 \leq I \leq P$), is sub-divided into a set of B_I sub-ranges, and the combinatorial product of all possible sub-ranges then defines the set of *bins*, or *cells*, that make up the partition. Each molecule is assigned to the cell that matches the set of binned characteristics for that molecule, and a subset is then obtained by selecting one (or some small number) of the molecules in each of the resulting groups. This procedure is summarised in Figure 8.

1. Select a set of *P* properties, and a set of B_I bins for each such property *I*.

2. Generate the $\prod B_I$ combinations of bins, each combination of which represents one of the groups in the partition.

3. Calculate the *P* property values for each molecule and allocate it to the appropriate group.

Figure 8. Algorithm for partition-based selection

The principal factor to be considered in partition-based selection is the set of characteristics that is used to define the chemical space. The first report of such an approach was by Mason *et al.* [58], who generated partitions defined by six global molecular properties that had been chosen to encode a molecule's hydrophobicity, polarity, hydrogen bond donor and acceptor power, torsional flexibility and shape. However, any sort of global property can be used to generate a partition, such as topological indices [59], BCUT parameters encoding atomic charges, atomic polarisabilities and atomic hydrogen-bonding abilities [60] and three-point pharmacophores [61]. Partition-based methods are discussed in detail elsewhere in this book and the reader is referred to the chapter by Mason [62] for further information on this increasingly popular approach to rational compound selection.

5. OPTIMISATION-BASED APPROACHES METHODS

All of the approaches described thus far are, in essence, trying to find a diverse (or, in the ideal case, the most diverse) subset from amongst the astronomically large number that can be generated from a database of non-trivial size. The algorithms to be discussed in this section reformulate the identification of the most diverse subset as a combinatorial optimisation problem.

The first such approach was described by Martin *et al.*, who used the theory of D-optimal designs in probably the first-ever paper on rational approaches to the selection of reagent libraries [63]. Although much cited in the literature, and described in detail in the chapter in this book by Andersson *et al.* [64], there have been only a few other applications of this approach, possibly because it tends to focus upon the selection of extreme outliers [29, 65]. Instead, we shall focus here on the use of genetic algorithms and simulated annealing since these, and other such approximate approaches to combinatorial optimisation, appear to be well suited to compound-selection applications.

As we have already noted, there is a very large, but finite, number of subsets that can be generated from a given dataset. Hence, if a diversity index is available that quantifies the degree of structural heterogeneity in a particular subset [47], a near-optimal subset can be obtained by exploring the space of all possible subsets to find that with the largest value of the chosen index, as summarised in Figure 9. This provides a simple, direct way of generating a diverse subset (and one that is clearly closely related to some of the maximum dissimilarity algorithms described previously). The first such approach was described by Gillet *et al.*, as part of a study to demonstrate that selecting subsets from fully-enumerated virtual libraries would yield more diverse sets of molecules than would selecting diverse sets of reagents [46]. Many of the experiments in this paper used a maximum-dissimilarity, *cherrypicking* algorithm, *i.e.*, one that selected individual molecules without regard to the synthetic efficiency of the resulting combinatorial library. This limitation was overcome by developing a *genetic algorithm* (or GA) to select a maximally diverse set of products that could be synthesised in a combinatorial manner whilst minimising the number of different reagents involved.

1. Initialise D to $-\infty$.

2. Generate the next size-n subset.

3. Update D if this is the most diverse subset thus far.

4. Return to Step 2 if there are still subsets to be tested.

Figure 9. Optimisation of a diversity index via exhaustive subset enumeration

A GA is a computational procedure that encodes potential solutions to the problem that is to be optimised in a data structure called a *chromosome*, and then processes populations of such chromosomes by means of *mutation* and *crossover* operators analogous to those encountered in Darwinian evolution [66, 67]. Each chromosome has an associated *fitness* score, which represents the extent to which that particular chromosome provides a solution to the problem that is to be investigated. Assume a two-component combinatorial synthesis in which n_1 of the N_1 possible first reagents are to be reacted with n_2 of the N_2 possible second reagents. The chromosome in their GA contains n_1+n_2 elements, each specifying one possible reagent, and the cross-product of these two sets of reagents then specifies one of the size-n_1n_2 possible combinatorial libraries that could be synthesised given the two complete sets of reagents. The fitness function for the GA is a diversity index quantifying the diversity of the size-n_1n_2 library encoded in each chromosome, and the GA thus tries to evolve chromosomes that maximise the value of this index. The index used by Gillet *et al.* was the mean pairwise dissimilarity (specifically the complement of the Tanimoto coefficient) when averaged over all the pairs of molecules in a size-n_1n_2 library, the molecules being represented by Daylight fingerprints [68]. This index is discussed by Pickett *et al.* [53] and Turner *et al.* [69] and was used here since it can be calculated very rapidly, a pre-requisite for use in a GA-based application where very large numbers of fitness values may need to be calculated. The selection algorithm has been extended since this initial report by the inclusion of several types of additional information in the fitness function, such as physicochemical property values and similarity to known drug molecules [70].

A very similar approach has been adopted by Hassan *et al.* [71] and by Agrafiotis [72], but here using a simulated annealing (SA) approach to explore the combinatorial search space. In SA, putative solutions are again generated and their fitness scores evaluated. A new solution is accepted if its fitness exceeds the best thus far; it is also accepted with a probability given by a Boltzmann distribution if its fitness is less than the current best score, thus preventing the search algorithm being trapped in a local minimum. The molecules are represented by principal components derived from calculated physical properties (topological and information content indices, and electronic, hydrophobic and steric descriptors) [71] or by low-dimensionality autocorrelation vectors describing the distribution of the electrostatic potential over the van der Waals' surface of a molecule [72], and the fitness scores are calculated using one of several different inter-molecular distance functions in the resulting descriptor space.

Another example of the use of SA as a searching tool is provided by the HARPick program developed by Good and Lewis [73]; a GA-based version has also been described [74]. Here, a molecule is characterised by its constituent three-point pharmacophores, these being generated from an approximate 3D structure, and the diversity of a set of molecules, such as a putative combinatorial library, is given by a function based on the number of distinct pharmacophores present in that particular library. The scoring function that drives the SA seeks not just to maximise the value of this index but also to ensure an approximately even distribution of a library's members across three properties that provide a crude, but rapidly computable, measure of molecular shape: these are the number of heavy atoms in a molecule, the largest triangle perimeter for any of the three-point pharmacophores in that molecule, and the largest triangle area for any of these pharmacophores. The ease with which additional factors can be included in a scoring function is an attractive feature of the GA and SA approaches (although there seems to be no real reason why such information could not be included in algorithms for the other approaches discussed in this chapter).

This chapter focuses upon selection algorithms, but it is perhaps worth mentioning that genetic algorithms, and related approaches to evolutionary computing, have been applied to several aspects of combinatorial chemistry: the reader is referred to the reviews by Clark for an extensive bibliography of work in this area [67, 75].

6. EVALUATION AND COMPARISON OF SELECTION METHODS

Cluster-based and dissimilarity-based methods for compound selection were first discussed in the Eighties but it is only in the last few years that the area has attracted substantial attention as a result of the need to provide a rational basis for the design of combinatorial libraries. The four previous sections have provided an overview of the main types of selection method that are already available, with further approaches continuing to appear in the literature. Given this array of possible techniques, it is appropriate to consider ways in which the various methods can be evaluated, both in absolute terms and when compared with each other. A method can be evaluated in terms of its *efficiency*, *i.e.*, the computational costs associated with its use, and its *effectiveness*, *i.e.*, the extent to which it achieves its aims. As we shall see, it is not immediately obvious how effectiveness should be quantified and we shall thus consider the question of efficiency first, focusing upon the normal algorithmic criteria of CPU time and storage requirements.

Thus far, we have discussed selection algorithms without much consideration as to the precise context in which they are to be used. However, the context will affect the choice of algorithm since those with large run-time and/or storage complexities can only be used when the selection task involves processing small numbers of molecules. The most obvious such application is the identification of a diverse set of molecules with some specific functionality to provide one of the sets of reagents in a combinatorial synthesis; another might be to choose, *e.g.*, a 10% sample from amongst a few thousands of analogues offered by a specialist supplier. Small numbers of molecules allow the use of selection procedures that would be infeasible otherwise; for example, the computational demands of clique-detection mean that the selection procedure described by Gardiner *et al.* [55] is limited to files containing a few thousands of molecules at most, and even this size of dataset would require very substantial computing resources. The great majority of the selection algorithms that have been described are not so limited, and many of them have, indeed, been developed specifically to handle the very large numbers of molecules that may be encountered in the processing of corporate databases and/or fully enumerated virtual libraries. The reader should note that we consider here just the selection step and ignore the computational requirements associated with the generation of the descriptors that are used as the basis for the diversity analysis: this can be the most demanding part of an analysis with certain types of descriptor, such as pharmacophore patterns where a conformational analysis must be carried out for each of the molecules that is to be considered [73].

Downs and Willett provide a detailed discussion of the computational complexities of the algorithms that have been suggested for cluster-based selection [21]. The stored-matrix algorithm for hierarchic agglomerative methods is only applicable to small files, as it has a complexity of $O(N^3)$ time and $O(N^2)$ space. The RNN version of Ward's method and the Jarvis-Patrick method (the two most widely used approaches) both have a complexity of $O(N^2)$ time and $O(N)$ space, thus allowing both methods to be used with files of up to about one-quarter of a million molecules on modern computer hardware; the Jarvis-Patrick method, which has a much lower constant of proportionality for the time complexity, can be applied to files up to perhaps five times larger. A straightforward implementation of the maximum dissimilarity algorithm in Figure 4 has a time complexity of $O(n^2N)$, which is actually $O(N^3)$ as n is typically some small constant fraction of N. However, $O(nN)$ algorithms have been described for the MaxMin [29, 42] and MaxSum [40] dissimilarity criteria, thus allowing these criteria to be implemented on a very large scale; for example, Higgs *et al.* state that a simulation of MaxMin-based selection of a ten-thousand molecule subset from a one-million molecule database took about 6 CPU hours on a SUN multiprocessor system [29]. It is difficult to carry out a complexity analysis for the sphere-exclusion algorithms without a probabilistic model for the distribution of inter-molecular dissimilarities, so that one can estimate the effect of the molecule-deletions in Step 3 of the algorithm shown in Figure 5; however, such algorithms are very quick in practice, especially if a random, rather than a deterministic, strategy is used to select the next molecule in Step 2 of the algorithm. Partition-based selection is very fast in execution, the existence of the binning scheme resulting in a time complexity of just $O(N)$, thus allowing even massive virtual libraries to be processed extremely rapidly. Finally, the combinatorial search spaces explored by the optimisation methods mean that they can run for extremely long periods of time. In fact, they seem to provide good solutions relatively quickly, with both Gillet *et al.* [46] and Good and Lewis [73] quoting run-times of a few tens of CPU minutes, on a Unix workstation with an R10000 processor, for the design of a combinatorial subset from a fully enumerated amide library. The main limitation with such approaches is the need for a very rapid fitness calculation, since this calculation must be repeated very many times during the identification of a subset, and for the prior enumeration of all of the members of the full virtual library, which can be demanding of both time and space if the library is large.

The simplest way to obtain a subset of a dataset is by means of simple random sampling but this is most unlikely to provide a subset that encompasses all of the structural classes present within that dataset. Instead, classes that are heavily populated in the dataset, such as the very large sets of analogues that characterise many corporate databases, will be represented proportionally in the subset, while low-frequency classes, where only a few molecules have been synthesised or otherwise acquired, are unlikely to be represented. Thus, while random selection samples the *molecules* that are present within a dataset, the selection methods discussed in this chapter are intended to sample the *classes of molecules* that are present within that dataset. The first question to be addressed when considering the effectiveness of the various methods is thus whether they do, in fact, perform better than random; only when an affirmative answer has been received to this question is it appropriate to consider which method (or class of methods) is the best of those that are available.

A statistical analysis of random and rational approaches suggests that there may not be much difference in effectiveness [76] and the results of early studies were sufficiently equivocal to provide some support for this idea. For example, one simulation study by Taylor [77] suggested that cluster-based selection was only marginally better than random selection at finding bioactive molecules in a dataset and that dissimilarity-based selection was worse than random; another simulation study by Lajiness, conversely, found that both cluster-based and dissimilarity-based selection were superior to random selection [50], and that the relative merits of the two approaches were determined in large part by the structural characteristics of the dataset that was being processed [78]. Cribbs *et al.* [65] compared D-optimal design, Ward's clustering and two partition-based methods for generating combinatorial libraries of ureas characterised by calculated physicochemical properties, and found that only the cluster-based approach appeared to be significantly better than random at covering the available 5D physical property space. van Geerestein *et al.* [79] found that cluster-based selection using Ward's method was noticeably superior to random selection only when small subsets (ideally 1-5% and certainly less than 25%) of the source dataset were employed, whereas both Brown and Martin [6] and Matter [8] have suggested that cluster-based selection is most effective with large subsets. More recently, Spencer has reported a large-scale retrospective analysis of high throughput screening results carried out at Pfizer, in which a maximum dissimilarity algorithm identified no more actives than did random selection [80]. It is increasingly accepted, however, that rational selection procedures *are* worth adopting [81] and we must then address the question of which types of procedure yield the best results, this use of the word 'best' implying the availability of some quantitative measure of selection effectiveness.

The effectiveness of selection methods can be evaluated on the basis of structural information of some sort, *e.g.*, by using the diversity indices discussed elsewhere in this book [47]. However, this is not fully appropriate in that while the selection methods described above seek to maximise diversity in *chemical space*, the principal reason for using such methods is to obtain subsets that maximise diversity in *biological space* [79, 82]. Thus, in the context of a general screening program one wishes to identify subsets that exhibit the widest possible range of different types of activity, while in a focused investigation one wishes to identify subsets that contain the largest number of molecules with the specific activity of interest. Indeed, if the biological assay provides quantitative, rather than qualitative, activity data one seeks subsets that exhibit the widest range of values in the chosen assay since that will provide the maximum amount of information for the derivation of a quantitative structure-activity relationship (QSAR); however, such data typically only become available once a lead discovery programme is well advanced.

Most approaches to the evaluation of effectiveness are based on the similar property principle of Johnson and Maggiora [10], which states that structurally similar molecules tend to exhibit similar activities. There is ample experimental evidence to support this principle [5-8, 20], which in turn implies that subsets exhibiting some degree of structural redundancy, *i.e.*, containing molecules that are near neighbours of each other, will also exhibit some degree of biological redundancy; a structurally diverse subset, conversely, should maximise the number of types of activity exhibited by its constituent molecules. It should thus be possible to compare the effectiveness of different structure-based selection methods by the extent to which they result in subsets that exhibit as many as possible of the types of activity present in the parent dataset. This simple, but appealing, approach to the comparison of *global diversity* procedures (to quote Matter and Lassen [83]) has been used by Matter [8] and by van Geerestein *et al.* [79] to compare different types of 2D and 3D molecular descriptors and by Snarey *et al.* [48] to compare different types of dissimilarity-based selection algorithms. The latter authors tested several maximum-dissimilarity and sphere-exclusion algorithms in experiments with a carefully selected sample of the World Drugs Index (WDI) database [84], and found that the best results were obtained with a MaxMin algorithm; a theoretical rationale for the superiority of this over other maximum-dissimilarity algorithms has been presented by Lobanov and Agrafiotis [13]. Snarey *et al.* also tested a more focused, *feedback* approach that considered just a single biological activity class. Here, the effectiveness of a subset was based on the number of active molecules not in the subset that were near neighbours of actives that were in the subset, *i.e.*, those additional actives that could be retrieved by a simple second-stage similarity search. The results from these additional experiments were much less clear-cut in that no single algorithm was consistently superior to the others [48].

Brown and Martin have described an extended comparison of methods for cluster-based selection [6, 7]. Their experiments used a 'leave-one-out' methodology, again based on the similar property principle, that was initially suggested by Adamson and Bush [85] and then first used on a large scale by Willett [20]: although used here for comparing clustering methods it would also be applicable to the comparison of different partition-based selection methods. The property value of a molecule, *I*, within a dataset is assumed to be unknown, and the classification resulting from the use of some particular clustering method is scanned to identify the cluster that contains the molecule *I*. The predicted property value for *I*, *P(I)*, is then set equal to the arithmetic mean of the observed property values of the other compounds in that cluster. This procedure results in the calculation of a *P(I)* value for each of the *N* structures in a dataset, and an overall figure of merit for the classification is then obtained by calculating the product moment correlation coefficient between the sets of *N* observed and predicted values. The most generally useful clustering methods will be those that give high correlation coefficients across as wide a range of datasets as possible; a related approach, but based on dissimilarities in property values, has been described by Patterson *et al.* for the evaluation of molecular descriptors [5]. In their experiments, Brown and Martin found that Ward's method was consistently superior to the other clustering methods that they tested, thus confirming previous, less detailed studies [20, 33].

Finally, most of the methods that have been developed thus far are empirical in nature, typically drawing upon existing algorithmic approaches, such as cluster analysis or evolutionary computing, and applying them to the chemical context. There is nothing wrong with such a pragmatic approach but it may result in the adoption of seemingly reasonable procedures that turn out, on closer inspection, to result in subsets with less than desirable characteristics. The need for a detailed set of comparative evaluation criteria was noted at an early stage in the development of methods for cluster analysis [86, 87] and Agrafiotis has shown that rigorous analysis can also be used to identify potentially ineffective methods in the present context; thus, he has used a geometric analysis to rationalise the observation of Snarey *et al.* [48] that the MaxMin maximum-dissimilarity method was notably superior to the very similar MaxSum method [13] and has also identified significant limitations in an information-theoretic diversity index originally described by Lin [88, 89].

7. CONCLUSIONS

Computer-based selection methods play an important role in the rational design of combinatorial chemistry programmes. However, the sheer range of methods that are already available means that it may appear difficult to identify which approach might be most appropriate for some particular application, as there are several factors that may need to be taken into account. Thus, considerations of computational complexity may be important if, *e.g.*, a large fraction of an entire corporate file is to be processed in a database-exchange application; different methods may be appropriate for, on the one hand, cherrypicking from an external database in a selective acquisition programme, as against, on the other, the specification of a combinatorial subset from a fully enumerated library design. The choice of method may also be influenced by the need to carry out other sorts of diversity analysis; for example, partition-based methods provide a simple and efficient mechanism not only for subset selection but also for database comparison.

This chapter has adopted a four-part classification of compound-selection methods, but other classifications are equally possible, such as whether the subset is identified directly or indirectly. Cluster-based approaches identify a set of diverse molecules indirectly, since the approaches require the initial identification of clusters of similar molecules, from which disparate molecules can subsequently be selected. The principal focus of interest is therefore the means by which the clusters are identified, with selection merely involving choosing one or some small number of compounds from each cluster in turn. Similar comments relate to the binning schemes that underlie partition-based approaches to compound selection; the principal focus of interest here is the representation that is used to generate the low-dimensionality space onto which the bins are mapped, and the act of selection is again trivial. The dissimilarity-based and optimisation-based methods are quite distinct from the first two classes in that the former both focus directly upon the identification of diverse subsets, without the need to create some intermediate data structure. Indeed, one might reasonably argue as to whether there are two distinct classes of method here; for example, Higgs *et al.* include both a MaxMin algorithm and a SAS optimisation routine in their discussion of what they refer to as *spread designs* [29].

An alternative two-part classification has been proposed by Pearlman *et al.* [90], who characterise methods as either *cell-based* or *distance-based*, these classes corresponding to partition-based methods and to all the other types of method, respectively. As Pearlman *et al.* note, distance-based methods can be used with any type of structural representation but are most effective when the need is to identify subsets (of whatever sort); cell-based methods can only be applied when low-dimensionality structure representations are available, but they are also very well suited to database-analysis and database-comparison tasks. The ease with which these latter tasks are accomplished by the two classes of method can be judged by comparing the studies reported by Shemetulskis *et al.* [34] and by Cummins *et al.* [59]. Given the computational efficiency and range of application of cell-based methods, they would appear to offer a very attractive approach to diversity analysis given the availability of appropriate low-dimensionality descriptors for the specification of the binning scheme. Many such descriptors have already been reported in the literature [62] and new approaches continue to appear, *e.g.*, the application of principal components analysis to high-dimensional fingerprint data [13].

Whatever subset-selection method is adopted, it should be used only after an initial filtering operation to remove from further consideration those molecules that exhibit some sort of undesirable characteristic, a process that Walters *et al.* memorably refer to as "REOS" (for "Rapid Elimination Of Swill") [91]. Examples of such characteristics include: the presence in a molecule of highly reactive or toxic substructures that have been catalogued in a corporate "badlist" of undesirable fragments (see, *e.g*, the partial list provided by Lajiness [38]); and restrictions on the values of properties such as the molecular weight, the octanol-water partition coefficient, and the numbers of rotatable bonds and chiral centres [92]. Filtering systems are increasingly being further refined by taking account of the "drug-like" nature of molecules, this information being obtained by similarity searching in databases of known drugs or by comparative statistical studies of databases of known drugs and of (presumed) non-drugs [93-95]; as noted previously, the latter type of information can be included in the fitness function of an optimisation-based selection procedure. Careful scheduling of the various components of a multi-level filtering system can result in the rapid elimination of a large fraction (of the order of 50% or more in some cases) of the molecules in a dataset at little computational cost, thus permitting more sophisticated types of analysis to be applied to the remaining members of the dataset.

In conclusion, the advent of combinatorial and HTS approaches to lead discovery has resulted in substantial interest in the development of novel techniques for computer-based compound-selection. This interest is being reflected not only in the application of novel algorithmic approaches to compound selection, *e.g.*, the use of k-D trees [13] and clique detection [55], but also in the increasing emphasis that is being placed on quantitative validation procedures, such as those discussed in the previous section; such developments can only further increase the importance of the methods discussed in this chapter.

ACKNOWLEDGEMENTS

I thank Dr Val Gillet for comments on this chapter and the Biotechnology and Biological Sciences Research Council, the Engineering and Physical Sciences Research Council, Glaxo Wellcome Research and Development, Pfizer Central Research, Tripos Inc., Tripos Receptor Research and Warner-Lambert Parke-Davis for funding my current research on computational methods for the analysis of molecular diversity. This paper is a contribution from the Krebs Institute for Biomolecular Research, which has been designated as a centre for biomolecular sciences by the Biotechnology and Biological Sciences Research Council.

REFERENCES

1. Barnard, J.M. Substructure searching methods: Old and new. *J. Chem. Inf. Comput. Sci.*, 1993, **33**, 532-538.
2. Downs, G.M. and Willett, P. Similarity searching in databases of chemical structures. *Rev. Comput. Chem.*, 1995, **7**, 1-66.
3. Good, A.C.and Mason, J.S. Three-Dimensional structure database searches. *Rev. Comput. Chem.* 1995, **7**, 67-117.
4. Martin. Y.C. and Willett, P., Eds. *Designing Bioactive Molecules: Three-Dimensional Techniques and Applications*; American Chemical Society: Washington, 1998.
5. Patterson, D.E., Cramer, R.D., Ferguson, A.M., Clark, R.D. and Weinberger, L.E. Neighbourhood behaviour: a useful concept for validation of "molecular diversity" descriptors. *J. Med. Chem.*, 1996, **39**, 3049-3059.
6. Brown, R.D. and Martin, Y.C. Use of structure-activity data to compare structure-based clustering methods and descriptors for use in compound selection. *J. Chem. Inf. Comput. Sci.*, 1996, **36**, 572-584.
7. Brown, R.D. and Martin, Y.C. The information content of 2D and 3D structural descriptors relevant to ligand-receptor binding. *J. Chem. Inf. Comput. Sci*, 1997, **37**, 1-9.
8. Matter, H. Selecting optimally diverse compounds from structure databases: a validation study of two-dimensional and three-dimensional molecular descriptors. *J. Med. Chem.*, 1997, **40**, 1219-1229.
9. Willett, P. Using computational tools to analyze molecular diversity. In *A Practical Guide to Combinatorial Chemistry*, Eds DeWitt, S.H. and Czarnik, A.W., Washington: American Chemical Society, 1997, pp17-48.
10. Johnson, M.A. and Maggiora, G.M. (Eds) *Concepts and Applications of Molecular Similarity*. New York, Wiley, 1990.
11. Dean, P.M. (Ed) *Molecular Similarity in Drug Design*. Glasgow, Chapman and Hall, 1995.
12. Willett, P. and Winterman, V. A comparison of some measures of inter-molecular structural similarity. *QSAR*, 1986, **5**, 18-25.
13. Lobanov, V. and Agrafiotis, D.K. A rational approach for combinatorial drug design. Paper presented at the Chemical Structure Association/Molecular Graphics and Modelling Conference on "Computational Approaches to the Design and Analysis of Combinatorial Libraries", University of Sheffield, 14-16 April 1998.

14. Brown, R.D. Descriptors for diversity analysis. *Perspect. Drug Disc. Des.*, 1997, **7/8**, 31-49.
15. Willett, P., Barnard, J. M. and Downs, G. M.. Chemical similarity searching. *J. Chem. Inf. Comput. Sci.*, 1998, **38**, 976-996.
16. The *Available Chemicals Directory* is distributed by MDL Information Systems Inc., 14600 Catalina Street, San Leandro, CA 94577, USA.
17. Sneath, P.H.A. and Sokal, R.R. *Numerical Taxonomy*; W.H. Freeman, San Francisco, 1973.
18. Everitt, B. S. *Cluster Analysis*; Edward Arnold, London, 1993.
19. Adamson, G.W. and Bawden, D. Comparison of hierarchical analysis techniques for automatic classification of chemical structures. *J. Chem. Inf. Comput. Sci.*, 1981, **21**, 204-209.
20. Willett, P., *Similarity and Clustering in Chemical Information Systems*, Research Studies Press, Letchworth, 1987.
21. Downs, G.M. and Willett, P. Clustering of chemical-structure databases for compound selection. In *Advanced Computer-Assisted Techniques in Drug Discovery*, Ed. van de Waterbeemd, H., 1994, New York, VCH, pp. 111-130.
22. Hodes, L. Clustering a large number of compounds. 1. Establishing the method on an initial sample. *J. Chem. Inf. Comput. Sci.* 1989, **29**, 66-71.
23. Whaley, R. and Hodes, L. Clustering a large number of compounds. 2. Using the Connection Machine. *J. Chem. Inf. Comput. Sci.* 1991, **31**, 345-347
24. Lance, G.N. and Williams, W.T. A general theory of classificatory sorting strategies. I. Hierarchical systems. *Comput. J.*, 1967, **9**, 373-380.
25. Murtagh, F. A survey of recent advances in hierarchical clustering algorithms. *Comput. J.* 1983, **26**, 354-359.
26. Ward, J. H. Hierarchical grouping to optimize an objective function. *J. Am. Stat. Assoc.*, 1963, **58**, 236-244.
27. Willett, P. An evaluation of relocation clustering algorithms for the automatic classification of chemical structures. *J. Chem. Inf. Comput. Sci.*, 1984, **24**, 29-33.
28. Forgy, E. Cluster analysis of multivariate data: efficiency versus interpretability of classifications. *Biometrics*, 1965, **21**, 768.
29. Higgs, R.E., Bemis, K.G., Watson, I.A. and Wikel, J.H. Experimental designs for selecting molecules from large chemical databases. *J. Chem. Inf. Comput. Sci.*, 1997, **37**, 861-870.
30. Jarvis, R.A. and Patrick, E.A. Clustering using a similarity measure based on shared nearest neighbours. *IEEE Trans. Comput.*, 1973, **C-22**, 1025-1034.
31. Menard, P.R., Lewis, R.A. and Mason, J.S. Rational screening set design and compound selection: cascaded clustering. *J. Chem. Inf. Comput. Sci.*, 1998, **38**, 497-505.
32. Willett, P., Winterman, V. and Bawden, D. Implementation of non-hierarchic cluster analysis methods in chemical information systems: selection of compounds for biological testing and clustering of substructure search output. *J. Chem. Inf. Comput. Sci.*, 1986, **26**, 109-118.
33. Downs, G.M., Willett, P. and Fisanick, W. Similarity searching and clustering of chemical-structure databases using molecular property data. *J. Chem. Inf. Comput. Sci.*, 1994, **34**, 1094-1102.
34. Shemetulskis, N.E., Dunbar, J.B., Dunbar, B.W., Moreland, D.W. and Humblet, C. Enhancing the diversity of a corporate database using chemical database clustering and analysis. *J. Comput.-Aid. Mol. Des.*, 1995, **9**, 407-416.

35. Doman, T.N., Cibulskis, J.M., Cibulskis, M.J., McCray, P.D. and Spangler, D.P. Algorithm5: A technique for fuzzy similarity clustering of chemical inventories. *J. Chem. Inf. Comput. Sci.*, 1996, **36**, 1195.
36. Nouwen, J. and Hansen, B. An investigation of clustering as a tool in quantitative structure-activity relationships (QSARs). *SAR and QSAR in Environmental Research*, 1995, **4**, 1-10.
37. Dunbar, J.B. Cluster-based selection. *Perspect. Drug Disc. Des.*, 1997, **7/8**, 51-63.
38. Lajiness, M.S. Dissimilarity-based compound selection techniques. *Perspect. Drug Disc. Des.*, 1997, **7/8**, 65-84.
39. Marengo, E. and Todeschini, R. A new algorithm for optimal, distance-based experimental design. *Chemometrics and Intelligent Laboratory Systems*, 1992, **16**, 37-44.
40. Holliday, J.D., Ranade, S.S. and Willett, P. A fast algorithm for selecting sets of dissimilar structures from large chemical databases. *QSAR*, 1995, **14**, 501-506.
41. Hudson, B.D., Hyde, R.M., Rahr, E. and Wood, J. Parameter based methods for compound selection from chemical databases. *QSAR*, 1996, **15**, 285-289.
42. Polinsky, A., Feinstein, R.D., Shi, S. and Kuki, A. LiBrain: software for automated design of exploratory and targeted combinatorial libraries. In *Molecular Diversity and Combinatorial Chemistry. Libraries and Drug Discovery*, Eds. Chaiken, I.M. and Janda, K.D., 1996, Washington: American Chemical Society, pp. 219-232.
43. *DiverseSolutions User's Manual.* St Louis MO: Tripos Inc., 1996.
44. Nilakantan, R., Bauman, N. and Haraki, K.S. Database diversity assessment: new ideas, concepts and tools. *J. Comput.-Aided Mol. Des.*, 1997, **11**, 447-452.
45. Clark, R.D. OptiSim: an extended dissimilarity selection method for finding diverse representative subsets. *J. Chem. Inf. Comput. Sci.*, 1997, **37**, 1181-188.
46. Gillet, V.J., Willett, P. and Bradshaw, J. The effectiveness of reactant pools for generating structurally diverse combinatorial libraries. *J. Chem. Inf. Comput. Sci* ., 1997, **37**, 731-740.
47. Gillet, V.J. Background theory of molecular diversity. In: *Molecular Diversity in Drug Design*, Ed. Dean P.M. and Lewis R.A., Kluwer, 1999, Ch. 3
48. Snarey, M., Terret, N.K., Willett, P. and Wilton, D.J. Comparison of algorithms for dissimilarity-based compound selection. *J. Mol. Graphics Modelling*, in the press.
49. Kennard, R.W. and Stone, L.A. Computer aided design of experiments. *Technometrics*, 1969, **11**, 137-148.
50. Lajiness, M.S. Molecular similarity-based methods for selecting compounds for screening. In*Computational Chemical Graph Theory*, Ed. Rouvray, D.H., 1990, New York: Nova Science Publishers, pp 299-316.
51. Bawden, D. Molecular dissimilarity in chemical information systems. In *Chemical Structures 2. The International Language of Chemistry*, Ed. Warr, W.A., 1993, Heidelberg: Springer-Verlag, pp. 383-388.
52. Holliday, J.D. and Willett, P. Definitions of 'dissimilarity' for dissimilarity-based compound selection. *J. Biomolecular Screening*, 1996, **1**, 145-151.
53. Pickett, S.D., Luttman, C., Guerin, V., Laoui, A. and James, E. DIVSEL and COMPLIB - strategies for the design and comparison of combinatorial libraries using pharmacophore descriptors. *J. Chem. Inf. Comput. Sci.*, 1998, **38**, 144-150.
54. Perry, N Selection of diverse database subsets by fingerprint and property-based methods. Paper presented at the Chemical Structure Association/Molecular Graphics and Modelling Conference on "Computational Approaches to the Design and Analysis of Combinatorial Libraries", University of Sheffield, 14-16 April 1998.

55. Gardiner, E.J., Holliday, J.D., Willett, P., Wilton, D.J. and Artymiuk, P.J. Selection of reagents for combinatorial synthesis using clique detection. *QSAR,* 1998, **17**, 232-236.
56. Babel, L., Finding maximum cliques in arbitrary and special graphs. *Computing,* 1991, **46**, 321-341.
57. Mason, J.S. and Pickett, S.D. Partition-based selection. *Perspect. Drug Disc. Des.*, 1997, **7/8**, 85-114.
58. Mason, J.S., McLay, I.M. and Lewis, R.A., In *New Perspectives in Drug Design,* Eds. Dean, P.M., Jolles, G. and Newton, C.G., 1994, Academic Press, London, pp. 225-253.
59. Cummins, D.J., Andrews, C.W., Bentley, J.A. and Cory, M. Molecular diversity in chemical databases: comparison of medicinal chemistry knowledge bases and databases of commercially available compounds. *J. Chem. Inf. Comput. Sci.*, 1996, **36**, 750-763.
60. Pearlman, R. S. "Novel software tools for addressing chemical diversity", accessible via WWW at URL http://www.awod.com/netsci/Issues/Jun96/feature1.html.
61. Pickett, S.D., Mason, J.S. and McLay, I.M. Diversity profiling and design using 3D pharmacophores: Pharmacophore-Derived Queries (PDQ). *J. Chem. Inf. Comput. Sci.*, 1996, **36**, 1214-1223.
62. Mason J.S. Absolute versus relative similarity and diversity. In: *Molecular Diversity in Drug Design*, Ed. Dean P.M. and Lewis R.A., Kluwer, 1999, Ch. 4.
63. Martin, E.J., Blaney, J.M., Siani, M.A., Spellmeyer, D.C., Wong, A.K. and Moos, W.H. Measuring diversity: experimental design of combinatorial libraries for drug discovery. *J. Med. Chem.*, 1995, **38**, 1431-1436.
64. Andersson, P.M., Linusson A., Wold S., Sjöström M., Lundstedt T. and Nordén, B. Design of small libraries for lead exploration. In: *Molecular Diversity in Drug Design*, Ed. Dean P.M. and Lewis R.A., Kluwer, 1999, Ch. 9.
65. Cribbs, C., Menius, A., Cummins, D.J., Scoffin, R. and Young, S.S., Paper presented at the 211th National Meeting of the American Chemical Society.
66. Devillers, J. (editor) *Genetic Algorithms in Molecular Modelling*. London: Academic Press, 1996.
67. Clark, D.E. and Westhead, D.R. Evolutionary algorithms in computer-aided molecular design. *J. Comput.-Aided Mol. Des.*, 1996, **10**, 337-358.
68. Daylight Chemical Information Systems Inc. 27401 Los Altos, Suite #370, Mission Viejo, CA 92691 USA
69. Turner, D.B., Tyrrell, S.M. and Willett, P. Rapid quantification of molecular diversity for selective database acquisition. *J. Chem. Inf. Comput. Sci.*, 1997, **37**, 18-22.
70. Gillet, V.J., Willett, P., Bradshaw, J. and Green, D. Selecting combinatorial libraries to optimise diversity and physical properties. *J. Chem. Inf. Comput. Sci.*, 1999, **39**, 169-177.
71. Hassan, M., Bielawski, J.P., Hempel, J.C. and Waldman, M., Optimization and visualization of molecular diversity of combinatorial libraries. *J. Comput.-Aid. Mol. Des.*, 1996, **10**, 64-74.
72. Agrafiotis, D.K. Stochastic algorithms for maximising molecular diversity. *J. Chem. Inf. Comput. Sci.*, 1997, 37, 841-851.
73. Good, A.C. and Lewis, R.A. New methodology for profiling combinatorial libraries and screening sets: cleaning up the design process with HARPick. *J. Med. Chem.*, 1997, **40**, 3926-3936.
74. Lewis, R.A., Good, A.C. and Pickett, S.D. Quantification of molecular similarity and its application to combinatorial chemistry. In *Computer-Assisted Lead Finding and* Optimization, Eds van de Waterbeemd, H., Testa, B. and Folkers, G., 1997, Wiley-VCH, Weinheim, pp. 137-155.

75. Clark, D.E. Evolutionary algorithms in computer-aided molecular design. At http://panizzi.shef.ac.uk/cisrg/links/ea_bib.html
76. Young, S.S., Farmen, M. and Rusinko, A. Random versus rational. Which is better for general compound screening? At http://www.netsci.org/science/screening/feature09.html.
77. Taylor, R. Simulation analysis of experimental design strategies for screening random compounds as potential new drugs and agrochemicals. *J. Chem. Inf. Comput. Sci.*, 1995, **35**, 59-67.
78. Lajiness, M. An evaluation of the performance of dissimilarity measures. In *QSAR: Rational Approaches to the Design of Bioactive Compounds,* Eds Silipo, C. and Vittoria, A., 1991, Elsevier Science Publishers, Amsterdam, pp. 201-204.
79. van Geerestein, V.J., Hamersma and van Helden, S.P. Exploiting molecular diversity: pharmacophore searching and compound clustering. In *Computer-Assisted Lead Finding and Optimization,* Eds van de Waterbeemd, H., Testa, B. and Folkers, G., 1997, Wiley-VCH, Weinheim, pp. 157-178.
80. Spencer, R.W. Diversity analysis in high throughput screening. *J. Biomolecular Screening*, 1997, **2**, 69-70.
81. Wikel, J.H. and Higgs, R.E. Applications of molecular diversity analysis in high throughput screening. *J. Biomolecular Screening*, 1997, **2**, 65-67.
82. Ferguson, A.M., Patterson, D.E., Garr, C.D. and Underiner, T.L. Designing chemical libraries for lead discovery. *J. Biomolecular Screening*, 1996, **1**, 65-73.
83. Matter H. and Lassen D. Compound libraries for lead discovery. *Chim Oggi*, 1996, 9-15.
84. The *World Drugs Index*, Derwent Information, URL http://www.derwent.co.uk/
85. Adamson, G.W. and Bush, J.A. A method for the automatic classification of chemical structures. *Information Storage and Retrieval*, 1973, **9**, 561-568.
86. Fisher, L. and van Ness, J. W. Admissible clustering procedures. *Biometrika,* 1971, **58**, 91-104.
87. Jardine, N. and Sibson, R. *Mathematical Taxonomy*. John Wiley, New York 1971.
88. Agrafiotis, D.K. On the use of information theory for assessing molecular diversity. *J. Chem. Inf. Comput. Sci.*, 1997, **37**, 576-580.
89. Lin, S.K. Molecular diversity assessment: logarithmic relations of information and species diversity and logarithmic relations of entropy and indistinguishability after rejection of Gibbs paradox of entropy mixing. *Molecules*, 1996, **1**, 57-67.
90. Pearlman, R.S., Smith, K.M. and Deanda, F. Low-dimensional chemistry spaces: recent advances. Paper presented at the Cambridge Healthtech Institute conference "Chemoinformatics" held in Boston 15-16 June 1998.
91. Walters, W.P., Stahl, M.T. and Murcko, M.A. Virtual screening - an overview. *Drug Disc. Today*, 1998, **3**, 160-178.
92. Lipinski, C.A., Lombardo, F., Dominy, B.W. and Feeney, P.J. Experimental and computational approaches to estimate solubility and permeability in drug discovery and development settings. *Adv. Drug Delivery Research*, 1997, **23**, 3-25.
93. Bemis, G.W. and Murcko, M.A. The properties of known drugs. 1. Molecular frameworks. *J. Med. Chem.*, 1996, **39**, 2887-2893.
94. Gillet, V.J., Willett, P. and Bradshaw, J. Identification of biological activity profiles using substructural analysis and genetic algorithms. *J. Chem. Inf. Comput. Sci.,* 1998, **38**, 165-179.
95. Sadowski, J. How to discriminate between drugs and non-drugs. Paper presented at the Chemical Structure Association/Molecular Graphics and Modelling Conference on "Computational Approaches to the Design and Analysis of Combinatorial Libraries", University of Sheffield, 14-16 April 1998.

Chapter 7

Molecular Diversity in Site-focused Libraries

Molecular diversity in site-focused libraries

DIANA C. ROE
Sandia National Labs, Mail Stop 9214, P.O. Box 969, Livermore, CA 94551

Key words: Site-focused libraries, structure-based combinatorial chemistry, structure-based design, 3D database searching, pharmacophore models, diversity

Abstract: This chapter examines different strategies for combining diversity and structure-based design in site-focused libraries. Site-focused libraries are libraries that concentrate on regions of diversity space containing specific chemical properties believed to be important for binding to a target receptor. These specific properties can be predicted from structure-based design. Combining this structure-based approach with combinatorial chemistry can provide a powerful tool to accelerate the drug design process. Diversity plays an important role developing these libraries. This chapter will begin with an overview of different strategies in structure-based design. It will go on to discuss several examples of how these strategies have been adapted to the design of large combinatorial libraries instead of single compounds. Several algorithms and example applications will be examined. The combiBUILD algorithm will be discussed in detail, as an example of how lead compounds can be rapidly identified by combining diversity with structure based design in site-focused libraries.

1. INTRODUCTION

An important first step in the design of a drug is to discover a lead compound that has both a promising therapeutic effect and pharmacological profile. Combinatorial chemistry coupled with high throughput screening has had a strong impact in the discovery and optimization of lead compounds. Typically 100-10,000 organic compounds can now be simultaneously synthesized and tested. Even so, this is far less than the number of compounds that could be potentially made, often in the range of billions or trillions. Therefore intelligent choices must be made. When no information is known about the target receptor, many assert that the best choice is to maximize the diversity of the compounds in the combinatorial library. However, it is not useful to explore regions in diverse space where the candidate ligands cannot be synthesized, or their pharmacokinetic and pharmacodynamic properties are poor. Thus library design should be directed towards relevant portions of diverse space. Similarly, when there is a lot of information known about the structural properties required of candidate ligands to bind to a target receptor, library design should be directed towards regions of diverse space which contain these structural properties. These libraries are often called 'site-focused' libraries, because they are focused towards candidate compounds containing certain structural properties at specific sites. However, diversity still plays an important role in developing these libraries.

Where can this information about the required structural properties come from? Computer-aided drug design provides specific predictions about which properties may be important in ligand binding. It has been shown to be very useful in both designing new ligands, and optimizing existing ones [1-3]. However, there are several problems due to inaccuracy in computational predictions. There may be errors of several orders of magnitude difference in predicting the binding affinity of a single ligand. Thus, any single prediction will only be correct a certain percentage of the time. This can be thought of as a casino black jack game. If you can count cards in the deck, you still are not guaranteed to win any particular hand. However, by the end of the week after playing thousands of hands you will almost certainly come home with more money than if playing randomly (assuming you're not kicked out first). Similarly, computer-aided design essentially provides the ability to "count cards". By combining computer-aided design with combinatorial chemistry, thousands of "hands" are played simultaneously, and the overall result should be much more effective than random. This new combined approach, which we call *structure-based combinatorial chemistry,* provides a powerful tool to identify new drug compounds rapidly.

There is an important interplay between structure-based combinatorial chemistry and diversity. By choosing the most diverse set of compounds within those predicted to bind well, more information becomes available about the structural properties required by compounds to bind, which refines and improves the predictions for the next round of synthesis. In addition, if good 'hits' are found, the information can be fed back to find compounds close in diverse space to the hit.

This chapter will begin with an overview of computer-aided drug design, focusing on structure-based approaches. It will then go on to describe how these techniques have been adapted to handle large combinatorial libraries instead of single compounds, giving a few examples to illustrate practical tools. Finally it will discuss several detailed examples of how lead compounds have been successfully found by integrating these two approaches.

2. COMPUTER-AIDED DRUG DESIGN OVERVIEW

Structure-based drug design was developed to take advantage of structural information of molecules to speed the process of finding and optimizing lead drug compounds [4-6]. As shown in figure 1, structure-based drug design is an iterative process, involving many different disciplines ranging from computer modeling to organic synthesis to biochemistry.

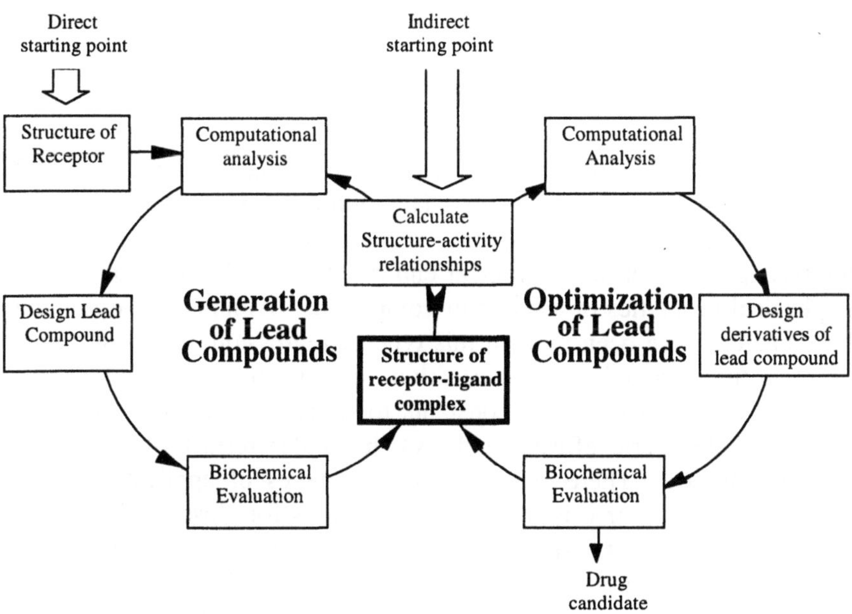

Figure 1. Overview of structure-based drug design

If the three dimensional structure of the target macromolecule is known by X-ray crystallography, NMR, or even homology modeling [7], this can serve as a starting point for what has been termed *direct* structure-based drug design [8]. This class of problems is usually referred to as receptor-based drug design. In this context, receptor is defined as any macromolecular target, and ligand as any inhibitor or substrate binding to the receptor. A molecular modeler examines and characterizes the three-dimensional structure of the target receptor and designs potential lead compounds predicted to bind to this target. These compounds are synthesized, or retrieved from a database, and evaluated for their biochemical effect. If a compound shows activity against the target, a new co-crystal structure may be solved of that compound binding to the target receptor. This information can be used both to improve the initial model and make better predictions for leads, and to go into a new cycle of improving the current lead.

In cases where the three-dimensional structure of the target is unknown, computational methods can still be applied using *indirect* design to infer information by examining a series of compounds with varying activities against the target receptor. From these compounds structure-activity relationships (SAR) between the compounds can be calculated, that is, how their two- and three-dimensional chemical structure relates to their observed binding affinities. Models derived from SAR data can be used to design new compounds, which feed back into the drug design cycle.

3. INDIRECT DESIGN

One of the first approaches developed for indirect design is called 2D-QSAR or the "Hansch" approach. This approach examines a congeneric series of compounds, and applies regression analysis to describe the relationships between the biological end point (e.g. inhibitory activity) and the physical properties of a series of compounds. Although a large number of parameters have been used in this type of regression, the most common are: (i) the octanol water partition coefficient as a measure of hydrophobicity; (ii) Hammett constants to describe electronic effects; (iii) Taft E_s for the steric effects; and (iv) the molar refractivity to describe dispersion forces. There are numerous examples of applications of this approach [9], however it is limited to predictions for closely related sets of compounds and cannot easily distinguish agonists from antagonists.

Another simple approach to finding new inhibitors based on SAR data is to search for compounds that are similar to existing inhibitors. Either two- or three-dimensional molecular descriptors can be calculated for each molecule, and a search is performed on a database to find compounds whose descriptors are most similar to the known inhibitors. Two-dimensional descriptors can be calculated very rapidly, allowing hundreds of thousands of structures to be processed in an hour. Three-dimensional descriptors are more challenging, since they require a time-consuming three-dimensional analysis of all molecule conformations, and it is suggested some two-dimensional descriptors may be superior to some of the three-dimensional ones [10, 11].

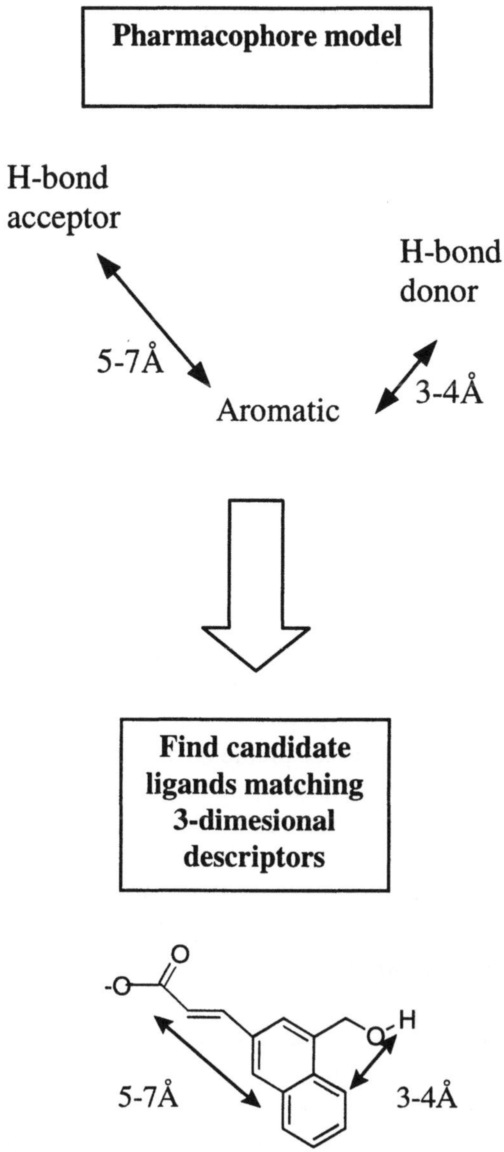

Figure 2. Discovering candidate ligands from a pharmacophore model. First the model is determined by examining SAR from a compound series. A database can then be searched for compounds matching the pharmacophore query

Another approach is to examine the common features of the active molecules and compare them to features in inactive molecules. Hydrogen bond acceptors and donors, positive and negative charged groups, hydrophobic groups, and aromatic groups are typical features. The set of features common to all active molecules is called the pharmacophore, which is an abstract model of all interactions believed to be important in binding. A three-dimensional pharmacophore also describes the relative three-dimensional orientation of these key interaction features relative to each other (see figure 2). A database can be searched for compounds that match the pharmacophore query, such as the program ALADDIN [12]. The challenges to this type of approach are generating the initial pharmacophore – a process known as pharmacophore mapping, and conformational sampling of the molecules in the database to match effectively the known pharmacophore. These methods have been reviewed in [13].

4. DIRECT DESIGN

The crux of receptor-based drug design is finding new compounds with specific binding properties. Although there are many different approaches to designing these new compounds, they can usually be broken down into three main steps (figures 2 and 3). First, a three-dimensional mapping of site points for key interactions is created. Second candidate molecules are found that can interact at these key site points. These compounds should be *complementary* to the target receptor. Finally the overall complex is evaluated by some function of its predicted binding affinity to the target receptor. Other properties, such as synthetic accessibility, bioavailability, low toxicity, etc., should be taken into account during this process, for design of ligands that are reasonable drug candidates. These steps are described below with example algorithms to illustrate each step.

4.1 Finding key site points

The first step in structure-based design is to identify key site points. These site points should represent places where a ligand atom or group could have a strong interaction with the receptor. There are several ways to characterize the receptor, depending on the interactions of interest. One of the simplest ways of analyzing the receptor is to examine its surface. The molecular surface is defined as the contact surface resulting from rolling a probe sphere (usually 1.4Å or the radius of a water molecule) along the surface of a protein [14, 15]. This will result in a van der Waals surface, where the edges or cusps between atoms have been smoothed out, representing the contact surface accessible to a water molecule. These surfaces are useful for describing shape complementarity, and can also be color-coded by electrostatic potential or hydrophobicity of the receptor, to provide some graphical energetic information.

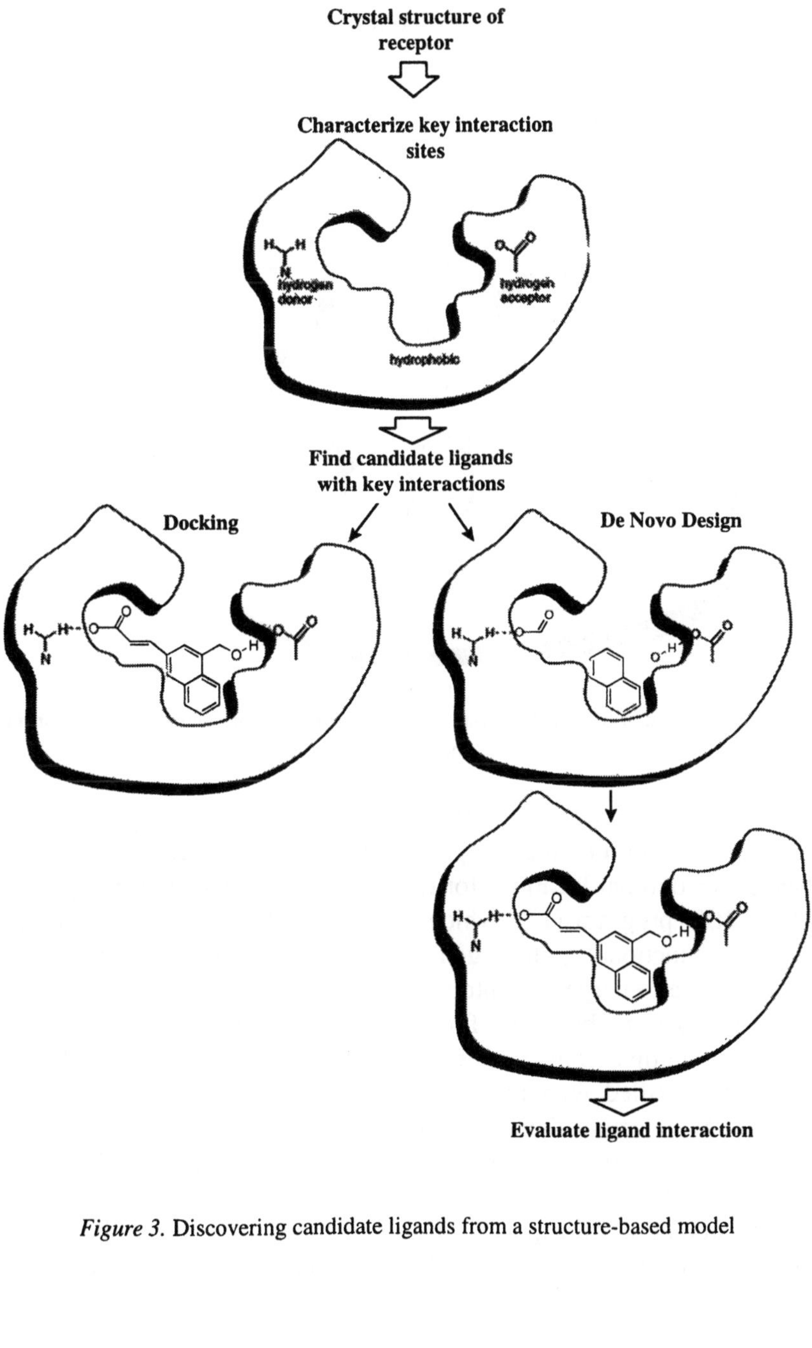

Figure 3. Discovering candidate ligands from a structure-based model

The surface can be used to determine shape-based site points. For example, the SPHGEN program from the DOCK program suite takes the molecular surface and fills all the grooves and pockets with spheres of varying size (see figure 4) to create a space filled "negative image" of the receptor [16]. This is useful for describing the shape complementarity required by the receptor, as a ligand should, in principle, fit within this set of spheres.

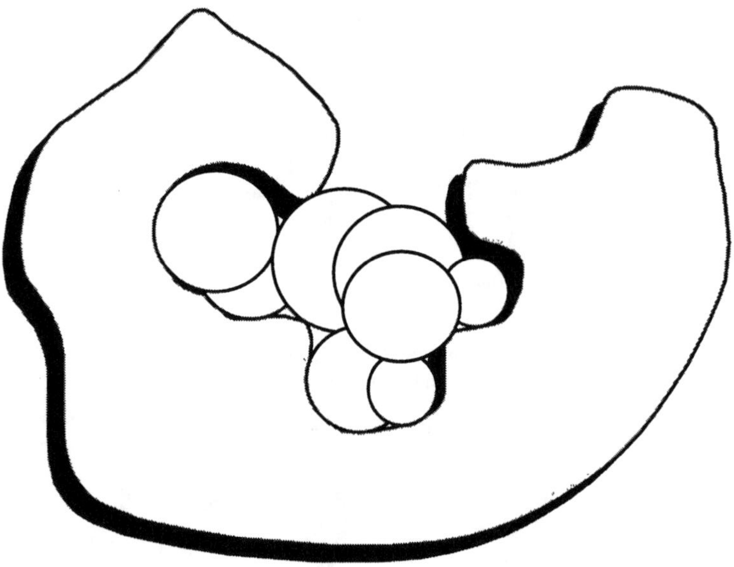

Figure 4. Example of SPHGEN output. The spheres define the 'negative space' for the receptor site

Other programs focus on energetic information. Several programs have been developed to probe the receptor for site points that represent energetic "hotspots", or optimal locations for ligand atoms. One of the first programs was the GRID program [17], which systematically probes the binding site with various functional groups such as amines, carboxylates, and hydroxyls. GRID creates an energetic contour map for each group, where the peaks represent interaction "hotspots" for that functional group. Another common approach is multiple-copy simultaneous search (MCSS) [18]. Here, the receptor site is filled with many copies of a functional group distributed randomly. A molecular dynamics simulation is run in such a way that the receptor interacts with all the functional groups simultaneously, but they do not interact with each other. Final orientations of the functional groups represent interaction "hotspots". These methods have been shown to be useful for generating site points that describe key binding interactions;

however they do not give a sense of the overall shape characteristics of the site.

Other programs use a rules-based approach to define site points. For example, several programs use geometric criteria to define site points for hydrogen-bonding interactions, derived from statistical analyses of hydrogen bonding distances and angles in X-ray crystallographic complexes [19]. The program LUDI uses these hydrogen-bond donor and acceptor site points, and additionally calculates lipophilic-aliphatic and lipophilic-aromatic site points, derived from a statistical analysis of non-bonded contacts in the crystal packing of small organic molecules [20]. Overall there is a variety of methods to describe the key site points.

4.2 Finding candidate molecules

Once the structure of the target receptor has been characterized, the next step is to find candidate molecules that bind to the target. This involves a search for compounds containing functional groups complementary to the key site points. There are two main approaches: database searching and *de novo* design (see figure 3). One database searching program is CAVEAT [21, 22], which uses vector/angular relationships between atoms as queries to search a three-dimensional database of small molecules. Post-processing examines if the molecules can sterically fit into the active site. The queries are often derived from known receptor/ligand complexes, where the goal is to mimic the key binding features of a known ligand while obtaining a novel scaffold. Another searching strategy is ALADDIN, which is based on pharmacophore searching, and was discussed earlier in indirect design. A third approach is the 'docking' strategy. These programs take a series of candidate ligands, dock each one into the target receptor, and try to predict its binding affinity. An example is the DOCK program shown in figure 5. Typically about 10-40% of DOCK compounds tested exhibit inhibition against a target receptor in the high micromolar range [23]. These programs have the advantage that they can be run on a database of existing compounds, which can be retrieved from inventory or purchased, rather than synthesized. This is also their limitation, as they are confined to existing compounds rather than novel scaffolds. Another difficulty is that the conformational space of each potential ligand is not well predicted, and so compounds can be missed because the correct conformer was never tested. Several methods have been developed to improve conformational sampling, either by docking multiple (usually 10-50) conformers of the same molecule [24], or by generating conformers on the fly during the docking run [25-27]. While these approaches can improve hit rates over single-conformer docking, they cannot yet be practically expanded to perform full conformational sampling over an entire database.

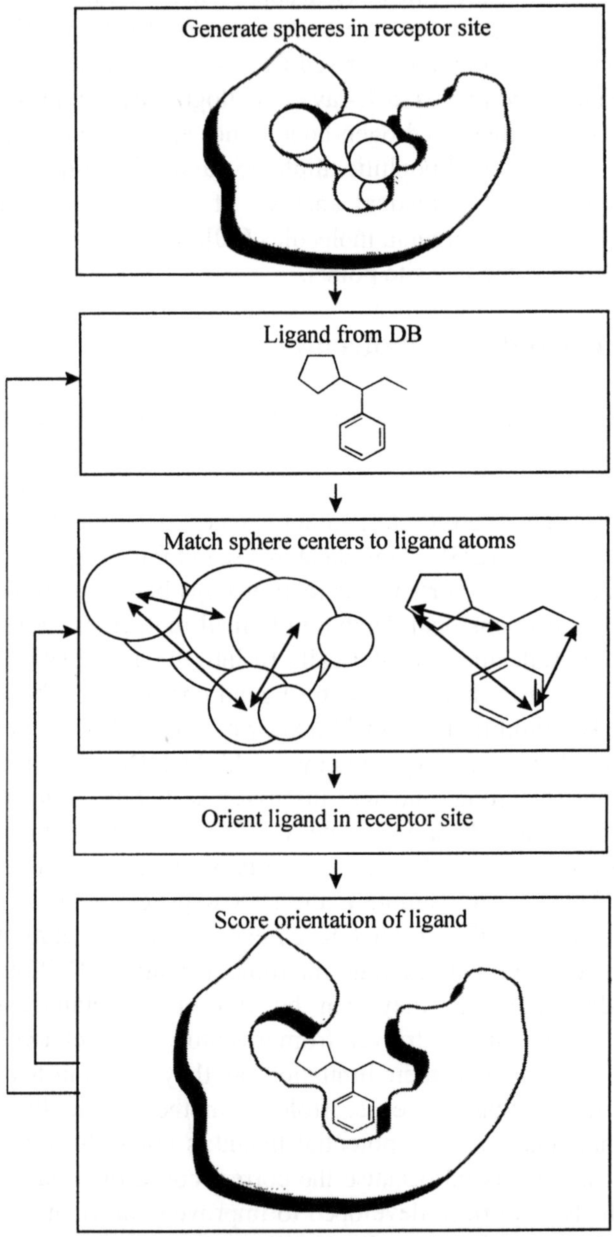

Figure 5. The DOCK algorithm. (1) A "negative image" is generated by filling the site with spheres. (2) a candidate ligand is retrieved from a database. (3) Internal distances are matched between a subset (usually 3-8) of sphere centers and ligand atoms. (4) The ligand is oriented

into the active site. (5) The interaction for that orientation is evaluated by a scoring function; the process is repeated for new orientations -- typically 10,000 orientations are generated per ligand. The top-scoring orientation is retained, and then the process is repeated for a new ligand.

In contrast *de novo* design attempts to build a compound up from scratch to fill the active site (see figure 3). This has the advantage that it is not limited to a set of compounds pre-existing in a database, and so completely novel scaffolds can be developed. Also, it allows searching over more conformational space, as the potential lead compound is built adaptively to best fit the site, and therefore is not limited to a set of pre-existing database conformations. However, new compounds may require significant time for synthesis. We developed the BUILDER program [28, 29], which starts with fragments in key binding areas from DOCK runs, and generates bridging groups on-the-fly from a molecular lattice, to generate a composite molecule. This entire process is interactive, so an organic chemist can guide the design to synthetically accessible compounds. Many other tactics have been taken for *de novo* design, ranging from growing atom-by-atom [30] or group-by group [31, 32] to form a new molecule, to other approaches to join fragments [33, 34]. Several successful *de novo* designs of ligands have been reported (for a review see Bohm [35])

4.3 Evaluating binding

Once a tentative ligand has been placed in a receptor, there needs to be some sort of scoring function to evaluate its possible binding energy. This has proved to be the most difficult task in structure-based drug design. The challenge is that a very fast scoring measure is needed, to rapidly evaluate not only a large number of compounds from a database, but also all the possible orientations and conformations of these compounds in the receptor. Another problem is accuracy. Although free energy perturbation and thermodynamic integration have been shown to be very effective in predicting differences in binding affinity between closely related ligands by "mutating" one compound into the other on the computer [36], it is still challenging to calculate absolute free energies of association. Even more restricting is that free energy calculations require long simulation times (of the order of weeks), which is not practical for scoring a database of compounds. Therefore much more approximate methods have to be employed.

One common approach is to use the intermolecular terms from the molecular mechanical force fields to evaluate the interaction energy. Such a potential usually includes terms for van der Waals and electrostatic interactions in the form:

$$F = \Sigma \{[A_{ij}/R_{ij}^{12} - B_{ij}/R^{6}_{ij}] + q_i q_j/\varepsilon R_{ij}\} \quad (1)$$

A_{ij}, B_{ij} are the nonbonded Lennard-Jones repulsion and attraction coefficients; R_{ij} is the interatomic distance between atoms i and j, and q_i and q_j are the atomic partial charges on atoms i and j; and ε is the effective dielectric constant. Sometimes solvation terms are added in, usually approximated from the solvent accessible surface area of molecules [37, 38]. Although this approach has been shown to correlate well with binding energy in a few examples [39], these correlations are system-specific and not generally transferable. More often, the error bars in affinity calculations are several orders of magnitude. This is not surprising considering that entropic contributions to energy and polarizability are completely neglected.

To combat the deficiencies in force-field evaluations, many people have turned to empirically derived methods. These assume that the receptor-ligand interactions are all additive, and they parameterize a number of terms including hydrogen bond interactions, charged interactions, hydrophobic interactions, and the number of rotatable bonds, by examining a number of protein ligand systems. A commonly used one is from the program LUDI, which was derived from examining the crystal structures of 45 protein-ligand complexes [40]:

$$\Delta G_{binding} = \Delta G_0 + \Delta G_{h\text{-}bond}\Sigma_{h\text{-}bonds}f(\Delta R,\Delta\alpha) + \Delta G_{ionic}\Sigma_{ionic}f(\Delta R,\Delta\alpha) + \Delta G_{lipo}\Sigma_{lipo}|A|_{lipo} + \Delta G_{rot}NROT \quad (2)$$

$f(\Delta R,\Delta\alpha)$ is a penalty function based on deviation from ideal distances (ΔR/Angstrom) and angles ($\Delta\alpha$/Degrees) in hydrogen bonds; it is assumed that the penalty function has a similar form for ionic interactions. $|A|_{lipo}$ is the contact surface between protein and ligand in Angstrom^2, and NROT is the number of rotatable bonds. This function fits the test data with a standard deviation of roughly 1.5 orders of magnitude.

Although attempts are made to create these functions to be generic for all systems, in practice they work best within the systems for which they have been parameterized and work less well when applied to novel receptor or ligand classes.

4.4 Current limitations

There are several pitfalls with the current structure-based drug design methodologies. The foremost is scoring, as yet there is no satisfactory potential that can accurately predict ligand-binding affinities on the fast time-scale needed to evaluate databases of potential drug compounds. The current scoring functions tend to have error bars of several orders of magnitude, which can roughly pick out potential binders from a database, but cannot easily differentiate between a nanomolar and micromolar inhibitor. Solvation, polarization and entropy are not treated well in the current methodologies. Another problem is the ligand conformation. Although several groups have developed schemes to examine ligand flexibility, it is computationally intractable to do a full ligand flexibility search on a large database of compounds. Similarly, almost all methods to date assume that the protein is rigid. Not only does this not take into account the small rotamer changes in protein sidechains that might result from ligand binding, but larger structural changes that may arise from induced fit in response to binding are neglected. Finally the issue of crystallographic waters or ions remains a challenge. Real ligands often have strong interactions with a water molecule or other ions, forming bridging interactions to the protein. Known crystallographic water molecules can be left in or out of the calculations. Although some groups have made progress in this area, there is no current technique to predict systematically and evaluate interactions with new water molecules.

5. CREATING SITE-FOCUSED COMBINATORIAL LIBRARIES

Computer-aided drug design programs can be adapted to help design combinatorial libraries. Combinatorial chemistry allows a large number of compounds to be synthesized simultaneously, essentially by generating variations on a theme. It starts with a central template or scaffold (usually a peptide-like backbone or heterocyclic ring system) that can be modified at several attachment sites. A set of reactants within the appropriate chemical class (such as amines, acylating agents, aldehydes etc) is chosen for each attachment site, and the exhaustive set of all combinations of reactants is synthesized in parallel. This can lead to potentially trillions of compounds. For example, consider a scaffold with 3 variable attachment sites, each with 8000 possible reagents. The size of the potential library would be 5.12×10^{11}compounds (see figure 6). Of course, many would be rejected in this case because they are not synthetically feasible. Even so, clearly more compounds exist than can be synthesized, so intelligent choices must be made.

Structure-based techniques can aid both in designing the initial scaffold for a library, and in choosing among various possible reactants for the library. To illustrate how these site-focused strategies can be combined with diversity to generate effective combinatorial libraries we will discuss some selective algorithms and examples in detail.

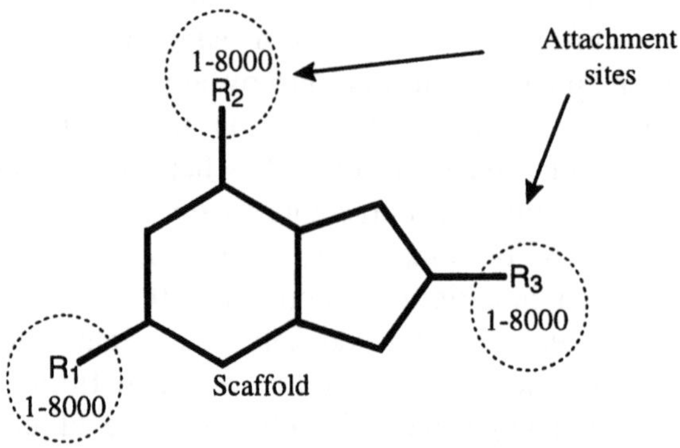

Figure 6. A schematic illustration of a combinatorial library design. In this library, there are three attachment sites on the scaffold. At each attachment site there are 8000 compounds that can be reacted, for a total of 8000X8000X8000 = $5.12X10^{11}$ potential compounds in this combinatorial library.

5.1 Scaffold Design

Many research groups start with scaffolds already known to be appropriate for the target receptor class, such as a transition state mimic, or a portion of an existing inhibitor. Structure-based techniques can be used to generate novel scaffold structures. When designing a scaffold, besides the standard considerations for maximizing the interaction with the target receptor, additional consideration needs to be taken for the positioning of attachment sites. Each attachment site should be in a location where its reactants can explore interesting regions of the target receptor, rather than in solution space, or worse, up against the edge of the receptor where no reactant can be attached. The same caveats exist with scaffold design as in traditional computer-aided drug design -- the predictions are approximate and only one can be tested at a time. Even so a few groups have successfully generated novel scaffolds [41].

An interesting twist is in the case where a designed scaffold is not amenable to efficient synthetic techniques. Graybil *et al* [42] handled that problem by designing a 'test scaffold'. They were looking for inhibitors of thrombin and designed an argatroban-like scaffold, which was a good lead compound but required an extensive synthetic effort. They generated a test scaffold based on the D-Phe-Pro-X dipeptide, which lends itself easily to combinatorial chemistry but is not suitable as a drug compound. They used this test scaffold to evaluate different sidechains in the S1 pocket of thrombin. The knowledge from the original argatroban scaffold, combined with docking experiments, was used to drive the design of the combinatorial library, and results from this SAR could then be used to help design modifications to their original argatroban-like scaffold. This approach generalizes to a design cycle with interplay between both a structure-based and a combinatorial scaffold series, as illustrated in figure 7.

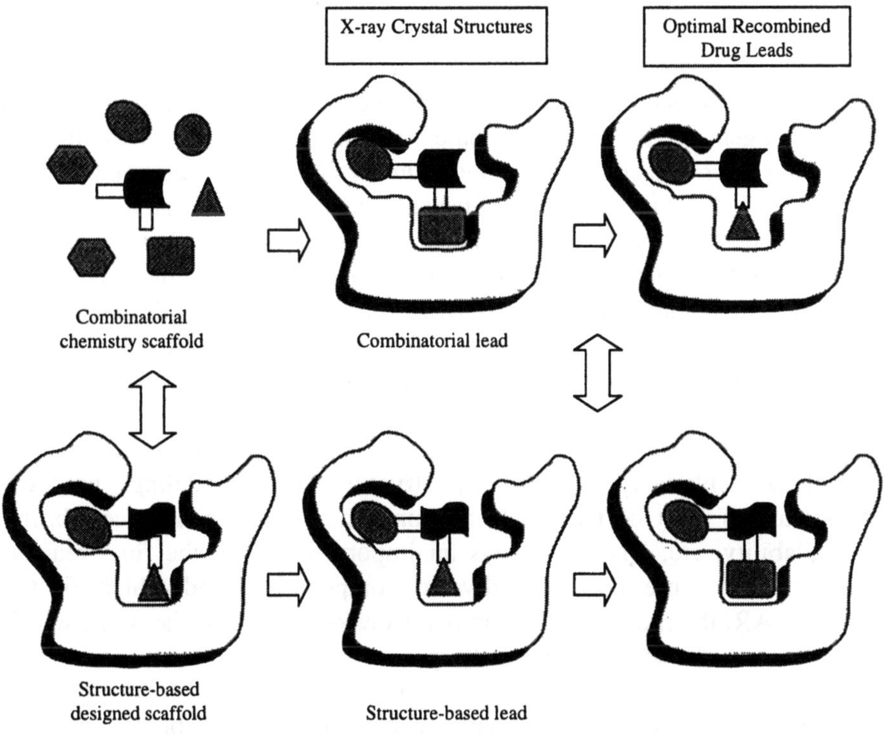

Figure 7. Generalized scheme for the interplay between a structure-based series and a combinatorial series of lead drug compounds

5.2 Virtual libraries

When structure-based strategies are used to help choose reactants for a combinatorial library, the strengths of both approaches complement each other. Although the structure-based design is not accurate, it statistically enhances and focuses the combinatorial library. Billions of possible library compounds can be generated on the computer in a "virtual library" and evaluated. From this, a smaller set of the order of thousands can be chosen for the actual library. In effect, thousands of predictions can be tested simultaneously, greatly enhancing the power of computer-aided design. Conversely, because the library was directed to regions of diverse space believed to contain crucial binding properties, the power of combinatorial chemistry is enhanced.

There are several new computational challenges presented by combinatorial libraries. First and foremost is how to handle the large number of compounds. For example, if the library of 5.12×10^{11} possible compounds from figure 6 were to be screened on the computer at the rate of 1 second/molecule, it would take over 16,000 years to screen all library members. Computational shortcuts must be taken. Several different approaches have been taken to handle this combinatorial explosion, such as applying a genetic algorithm to sample compound space [43], or by initially examining each attachment site individually, to screen out uninteresting reactants [44, 45]. Once a virtual library has been evaluated, and a set of top-scoring compounds has been determined, the next concern is how to use this information to select the optimum set of reactants. In other words, to calculate which subset of reactants will produce a library with the greatest number of top-scoring compounds. This can be handled directly, by optimizing the entire set of compounds generated from a subset of the reagents, or indirectly by examining final results and choosing reactants with the greatest frequency in the top scoring compounds. Finally, after reagents have been tentatively selected by structure-based modeling, practical considerations must be taken into account, such as synthetic accessibility, bioavailability, etc. Diversity plays an important role in the final reagent selection. The more the more diverse compounds tested which fit the existing SAR, the more information is retrieved to improve the SAR, which further focuses the search in diversity space.

The next sections describe several specific examples of how the full range of computational methods have been used to design libraries, ranging from methods using only 2-D structural descriptors, to complete structure-based combinatorial design.

5.3 Incorporating SAR data into library design

One of the original approaches for choosing reactants for combinatorial libraries was by Sheridan *et al.* [43], who used a genetic algorithm to direct library design towards the known SAR data. In an example they examined a library made up of tripeptoids. Peptoids are polymers of N-substituted glycine (figure 8), which are generated by joining amines with linker fragments. As there were several thousand possible primary and secondary amines available, and three amine attachment sites in a tripeptoid, there were over 20 billion possible compounds in this library. The goal was to find library compounds that matched the existing SAR data (i.e. were chemically similar to known active compounds). In this example they used two-dimensional topological descriptors to measure similarity and score the library compounds, but the overall approach could also be generalized to many other two-dimensional or three-dimensional descriptor sets. The genetic algorithm began with an initial population of 300 compounds with randomly selected amines at each attachment site, which was slowly modified through mutation and crossover, until the average score of the population converged (typically < 25 generations). This strategy was tested using several peptide lead compounds as input, and was found to return very similar peptoid counterparts. Although they did not test the resulting libraries experimentally, their experience with database searching had shown that these topological descriptors could increase the 'hit rate' by 5- to 40-fold compared to random screening.

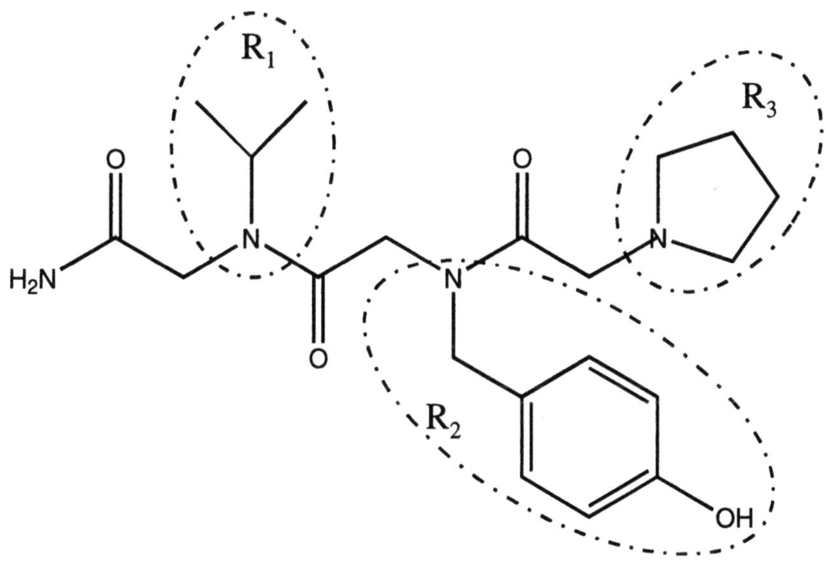

Figure 8. A library of tripeptoids. Peptoids are polymers of N-substituted glycine. This library has three attachment sites (R1-R3) for primary or secondary amines

One difficulty in applying this type of algorithm to combinatorial chemistry is how to choose the best compounds from the final population. It is not possible to synthesize the entire final population, since it consists of random combinations of *all* amines, rather than a systematic combination of a *selected set* of amines, as required for combinatorial synthesis. This is handled by choosing the amines with the highest frequency in the final population to be in the selected set of reactants. Another issue is that since genetic algorithms are stochastic in nature, they can converge to different solutions when given different starting seeds. It is therefore important to try several different starting seeds, and generate results from the combination of all runs.

5.3.1 Designing a library from a pharmacophore model

Although using two-dimensional SAR information has been demonstrated to be an effective tool in predicting binding, real compounds bind in three-dimensional space. Approaches to choosing reactants for combinatorial libraries using three-dimensional SAR information involve performing a three-dimensional database search. An example is the program DANTE [46-48], which analyzes the SAR data to generate a shape-enhanced pharmacophore model, which can be used to guide a three-dimensional search through a list of potential library reagents.

DANTE begins by identifying a common pharmacophore for active compounds using the most-selective heuristic (see figure 9). This is an iterative procedure, beginning with all pharmacophore dyads (i.e. pharmacophore subsets consisting of two functional groups and the distance between them), and ranking them by their selectivity, which is a measure of likelihood of finding that pharmacophore in the set of actives compared to finding it in a set of random compounds. Next all triads (subsets of 3 functional groups) which contain the most selective dyad are ranked, then tetrads, etc until the selectivity is no longer improved.

Once the pharmacophore is generated, the active compounds are superimposed and the "shrink-wrap" surface, defined as the surface of smallest volume that encloses at least one conformer of each active molecule, is calculated. This surface can be analyzed to provide raw shape information (figure 10). Regions of the shrink-wrap surface that are penetrated by inactives that otherwise match the pharmacophore can be marked as "forbidden regions". Other regions which haven't been explored can be marked as "terra incognita", or unknown regions.

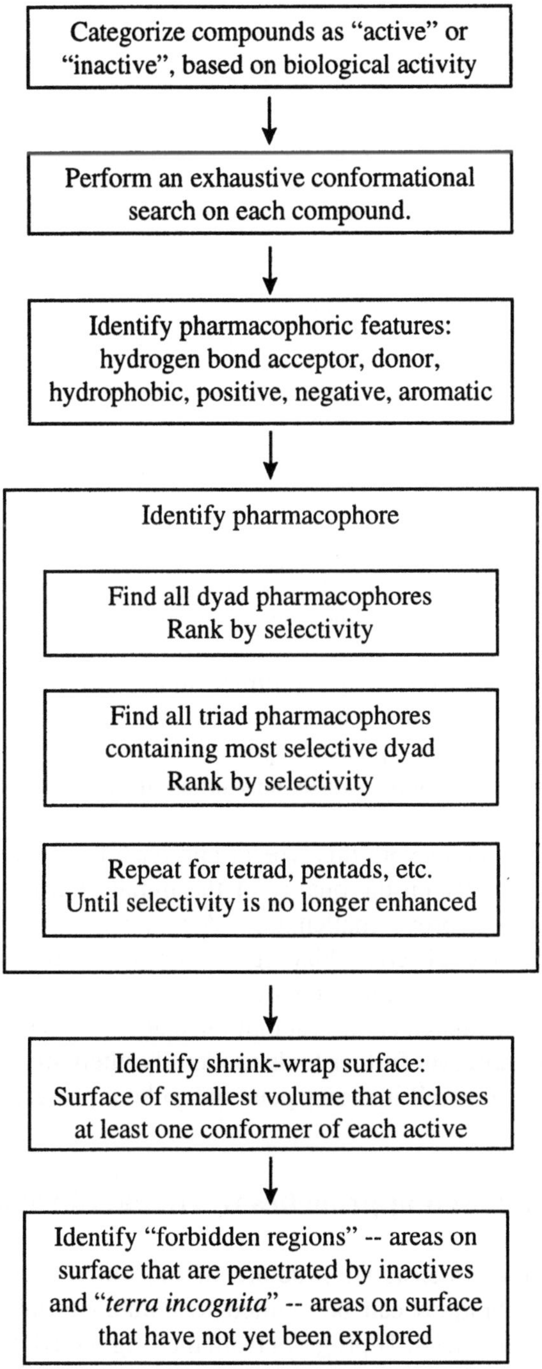

Figure 9. The DANTE methodology

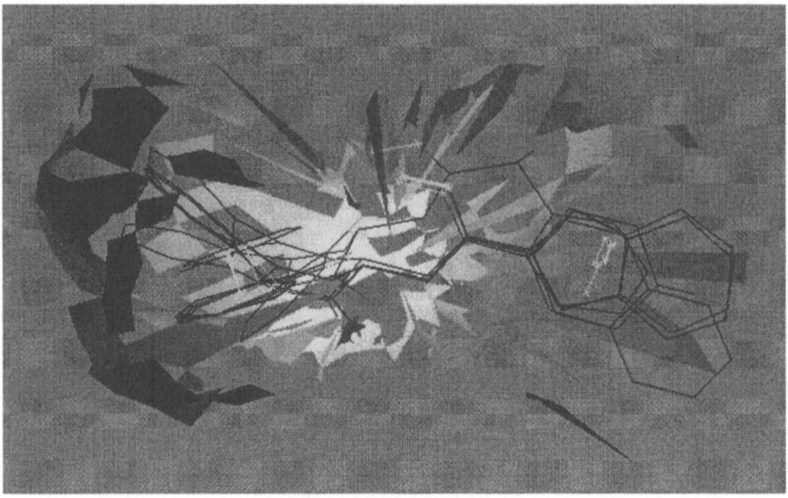

Figure 10. An example of a shape-enhanced pharmacophore for a series of benzodiazepines as CCK antagonists. "Terra incognita" is shown transparently and "forbidden regions" are shown as opaque

This model can guide the three-dimensional search through potential library reagents by directing reagent search towards possible sites of good interaction (i.e. towards the pharmacophore), away from bad interactions in the "forbidden regions", and towards new areas of diversity to explore in the "terra incognita".

One of the difficulties with this type of approach is that the quality of the results depends directly on the quality of the initial pharmacophore model, which can sometimes prove challenging. A particular hazard is that active compounds can have different binding modes, even in a closely related series of inhibitors. This algorithm partially addresses that problem, since it does not take as a fundamental assumption that all molecules bind in the same orientation, however further tools to identify alternative binding modes would be helpful. Overall it shows potential to be a powerful tool to direct library design.

5.4 Structure-based approaches to virtual libraries

When the three-dimensional structure of the target receptor is known, receptor-based techniques can be applied to combinatorial library design. We adapted the combiBUILD program from the original BUILDER program to generate and evaluate these virtual combinatorial libraries. It was developed for the case where the *orientation* of the scaffold in the active site is known, and therefore focuses on examining the *conformation*s of the possible components that can be attached to the scaffold. A similar program developed simultaneously is PRO_SELECT, which was applied to finding inhibitors of thrombin [45, 49]. We used combiBUILD to find inhibitors for cathepsin D, which we will discuss in detail to illustrate how rapidly site-focused libraries, combined with diversity, can lead to the design of nanomolar inhibitors.

The combiBUILD algorithm starts with a scaffold with a known orientation in the active site of the target macromolecule. It attaches each possible component to the scaffold, performs a systematic conformational search, and evaluates each component conformation within the active site with a force field score (see figure 11). One challenge with virtual libraries is how to reduce the combinatoric problem of examining every possible conformation of every component at every attachment site. CombiBUILD handles this challenge by examining each attachment site independently of the others. This turns the expensive multiplicative problem of testing all possible combinations of components, into a simple additive problem of testing all individual components (see table 1). However, this leads to the problem of how to handle components that can interact and clash with each other. We employed a probability-based clash grid, which allows the components to affect each other is an averaged way, without resorting to a time-consuming combinatoric search. A side benefit of treating each attachment site independently is that the final results provide direct predictions for the set of reactants to choose, rather than having to look at their frequencies in a final set of compounds.

Table 4. Testing all possible combinations of components vs. testing all individual components

	All Possible Combinations of Components	All possible Individual Components
Components	$[R_1] \times [R_2] \times [R_3]$	$[R_1] + [R_2] + [R_3]$
Number of Components	700x1900x1900	700+1900+1900
Total number of compounds to examine	2.5 billion	4500

We tested combiBUILD by using it to help design inhibitors for cathepsin D. Cathepsin D is an aspartyl protease whose structure has been solved with the inhibitor pepstatin. It is implicated in several disease processes including Alzheimer's amyloid plaque formation and breast cancer metastasis. Our goal was to design a small library of 1000 compounds of possible inhibitors to Cathepsin D.

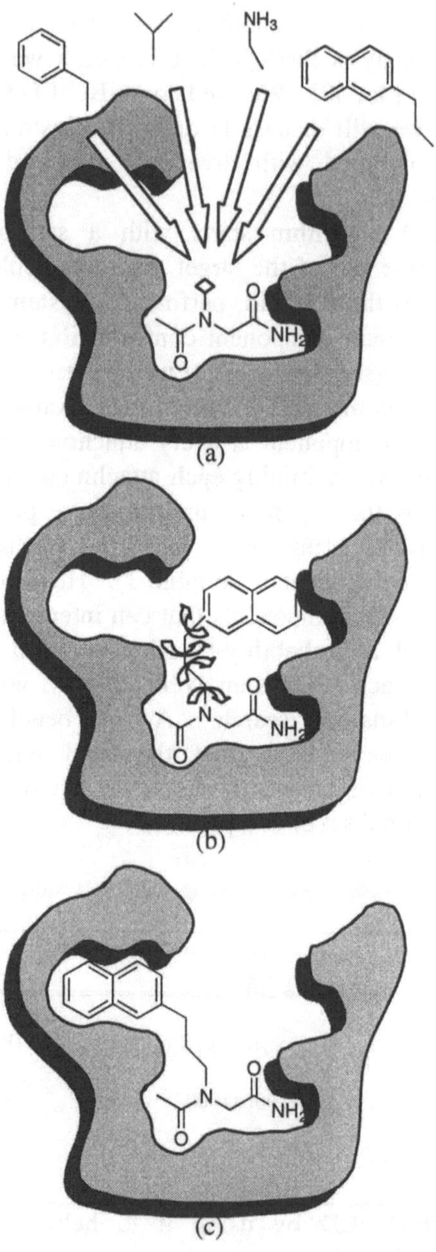

Figure 11. The combiBUILD algorithm: (a) Starting point is a scaffold modeled into its target receptor. CombiBUILD goes over a database of possible reactants to add to the scaffold; (b) CombiBUILD performs a systematic rotational search on each reactant (typically in 10-15 increments within the low energy minima for each torsion type), pruning for intramolecular clashes and bumps to make the search more efficient; (c) CombiBUILD performs a rigid body minimization on each possible component conformation, scores it with a force-field scheme.

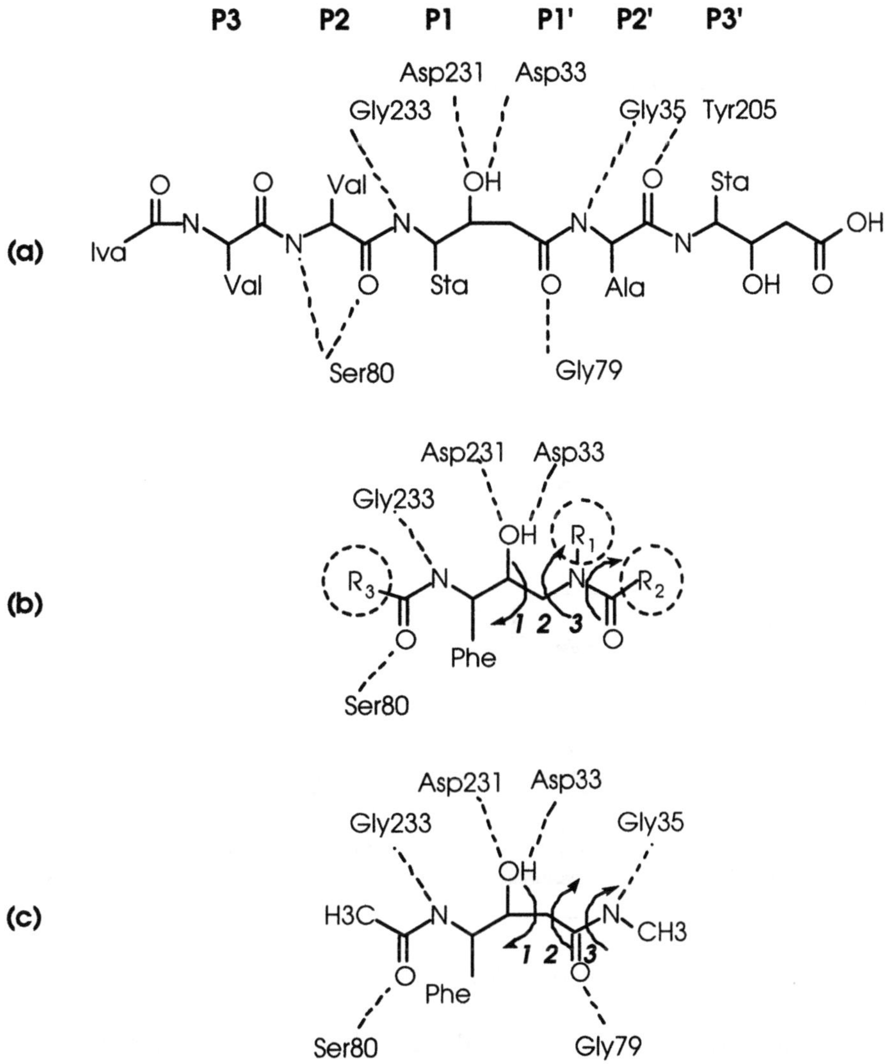

Figure 12. Combinatorial Library Scaffold compared to pepstatin. (a) Pepstatin (b) scaffold for combinatorial library (c) portion of pepstatin used to test methodology. The P1-P3' peptide side-chains are indicated

A synthetic scheme was devised for inhibitors of cathepsin D based on pepstatin, however using a (hydroxyethyl) amine isostere as the tetrahedral mimic instead of statine. A comparison of this scaffold to pepstatin is shown in figure 12. The synthetic procedure required three components, aside from the scaffold, constituting the R_1–R_3 groups (Figure 13). Candidate components were found by searching the ACD [50], a database of commercially available compounds, for ones with the correct reactant properties. There were a total of over 1 billion possible compounds that could be synthesized, from which we needed to choose 1000.

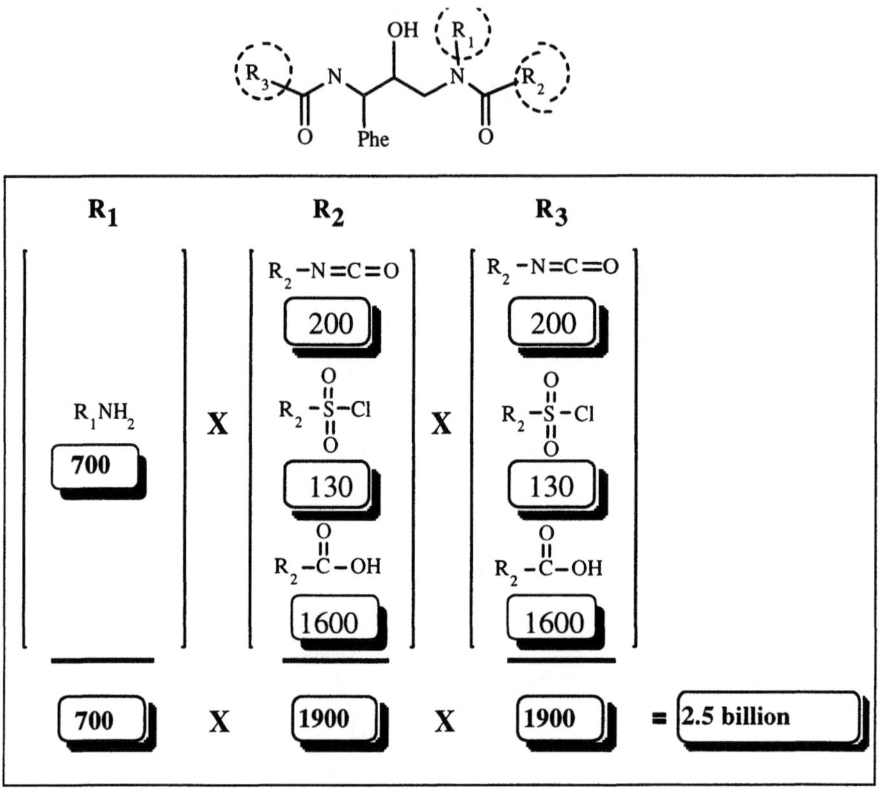

Figure 13. Combinations of Components of the Combinatorial Library

As combiBUILD assumes a fixed scaffold in the active site as its initial premise, our first job was to model the conformation of the scaffold. The conformation of the pepstatin inhibitor was used as a starting point for predicting the scaffold conformation. In figure 12, the scaffold is identical to pepstatin on the P1-P3 side, and therefore is likely to adopt the same conformation to maintain the hydrogen bonds to Ser80, Gly233, and the active site aspartyls. However, the P1'-P3' side differs from pepstatin, and cannot form hydrogen bonds to Gly 79 and Gly35. Therefore, we chose to rotate systematically the three torsion angles on this side, and minimize them using the AMBER force field [51]. The resulting minimized conformations were clustered into four similar families of conformations

Finally, we were ready to try the combiBUILD program. CombiBUILD placed each of the R_1, R_2, and R_3 components onto our modeled scaffold in the protein active site and performed a full conformational search of each. All components with greater than 4 rotatable torsions were pre-screened and removed. This is a reasonable choice, since these components would probably have conformational entropies that are too large to be interesting choices. To handle possible interactions between the R_1 and R_2 components, we used a probability grid that was generated from the top 50 results of a preliminary combiBUILD run. Figure 14 shows how the R_1 and R_2 components in this preliminary run had a lot of overlaps. The R_1 components were used to generate a clash grid for R_2 and vice-versa. The grid was set up so that on average if an R_1/R_2 conformation clashed with more than 50% of the R_2/R_1 components, that conformation was discarded. Figure 15 shows the top results when the clash grid was applied to both the R_1 and R_2 components.

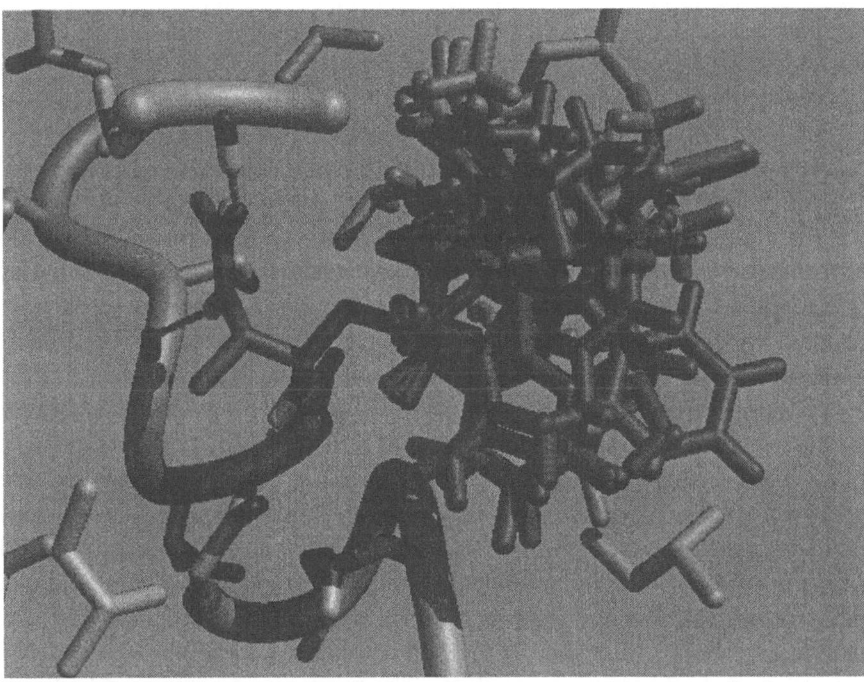

Figure 14. BUILDER conformations of R1 and R2 components without a clash grid

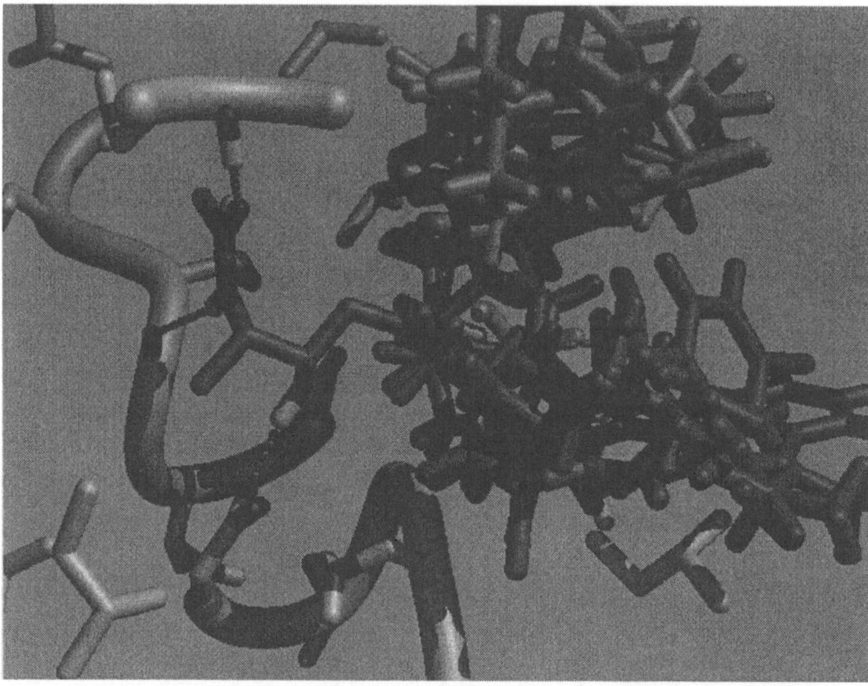

Figure 15. BUILDER conformations of R1 and R2 components using a clash grid

This entire strategy was applied to each of the four possible scaffold conformations. The top 50 components for each of the scaffold families were merged for each attachment point (i.e. R_1, R_2, R_3), and hierarchically clustered to maximize the diversity of the compounds chosen for library synthesis. In the end, the 10 best scoring compounds from unique clusters were chosen for each R_1, R_2, R_3 attachment point. Thus, our final library had the most diverse set of components that fitted our structural hypothesis.

To evaluate the results of the directed library approach, a purely diverse library of 1000 compounds was also generated as a control. The overall results are shown in figure 16, comparing the number of compounds found in the directed library to the number of compounds found in diversity alone. Not only did the directed approach yield more hits, but also the ratio of directed/diverse hits increased as the potency increased (going from roughly twice as many to 7 times as many hits. The most potent compound had a K_i of 73nM for the directed approach.

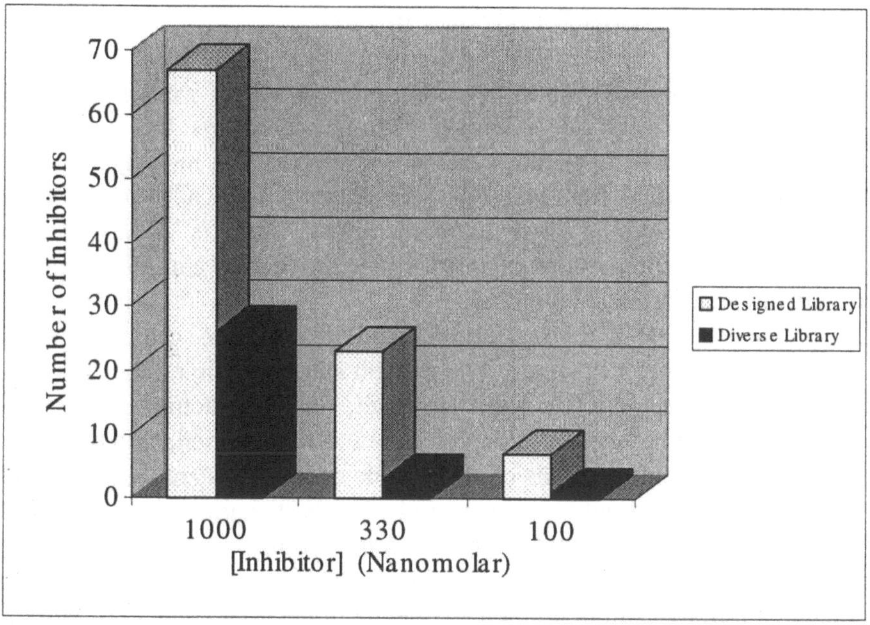

Figure 16. Number of compounds inhibiting cathepsin D by concentration of inhibitor. Result of high through-put fluorescence assay for cathepsin D

The diversity clustering analysis that was used to choose the final components of the directed library proved useful to design a second generation library quickly. We re-examined the clusters of the top scoring components, and generated a small library of 39 compounds from other components in those clusters. These are components that combiBUILD predicted would bind well, and which were similar to compounds that had already been shown experimentally to bind well. Almost all (92%) had IC_{50} better than 1μM, and 18% were nanomolar inhibitors, the best with a K_i of 9nM.

Thus by using structure-based drug design to develop a site-focused library, combined with diversity analysis of the top predictions, we were able to develop potent compounds quickly. There are several improvements that could be made to the combiBUILD algorithm, such as employing an intramolecular term to evaluate interactions, adding an estimated entropy term for the number of rotatable bonds, and using a more efficient conformational search which does not increase combinatorially with the number of bonds. None the less, this example shows the power of how a simple structure-based approach can greatly improve potency in the final library design.

5.5 Other programs for virtual libraries

Although combiBUILD has been shown to be useful when the binding orientation of the scaffold is known, that is not always the case. The combinatorial DOCK algorithm has been developed as a more general tool for examining virtual libraries [52]. Combinatorial DOCK places more emphasis on searching *orientational* space, that is how possible combinatorial compounds are oriented in the active site, and less emphasis on searching conformational space.

Combinatorial DOCK is a simple variation on the original DOCK algorithm (see figure 17). It also begins with spheres defining the "negative space" of the receptor active site. However, for the matching step it only uses atoms on the library scaffold. Once the scaffold is oriented in the active site, a library of conformations of all possible components are placed at each attachment point, and individually scored. The best scoring components are then examined in combination and tested for intramolecular clashes. The procedure is repeated for new orientations. The best set of components is initially chosen by examining their frequency in the final results, as with the genetic algorithm. Further optimizations are then performed examining combinations of components.

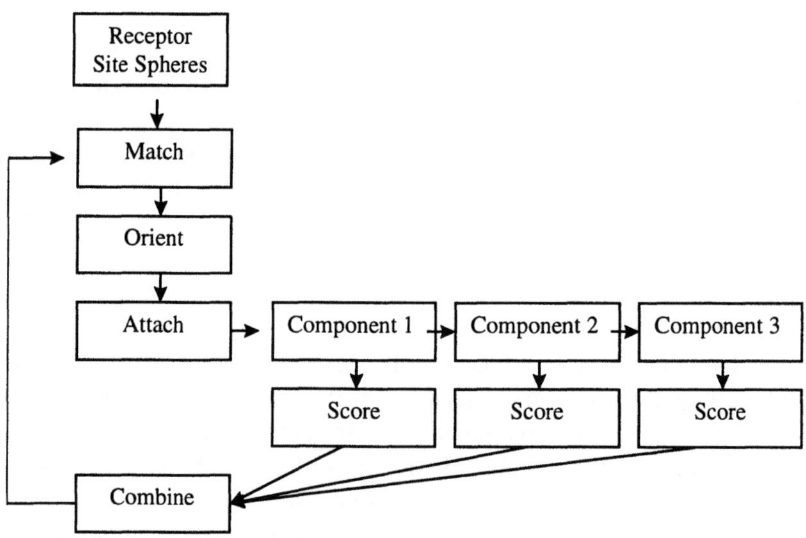

Figure 17. The combinatorial DOCK algorithm. The receptor site is filled with spheres and then the scaffold atom-atom internal distances are matched to the site sphere–sphere distances. The matching process is used to orient the scaffold inside the active site. All components from all attachment sites are placed on the scaffold individually and scored. The top scoring components are combined on the scaffold, and tested for intramolecular clashes. Resulting best scores are saved. The process is repeated for a new orientation.

This approach was verified by performing a retrospective examination of the library designed by combiBUILD for cathepsin D. Since experimental results exist for all the library compounds, the question was how well could the combinatorial DOCK algorithm predict the active compounds from this set, compared to random. This was measured by looking at the "enrichment factor" or the ratio of active compounds predicted by DOCK compared with those expected from random predictions (see figure 18). For example, using 330nM to define active, DOCK finds approximately 78% of actives in the first 75 compounds (out of 1000 compounds total) and 87% of the actives in the top 200 compounds, for an overall enrichment factor of 10 [53].

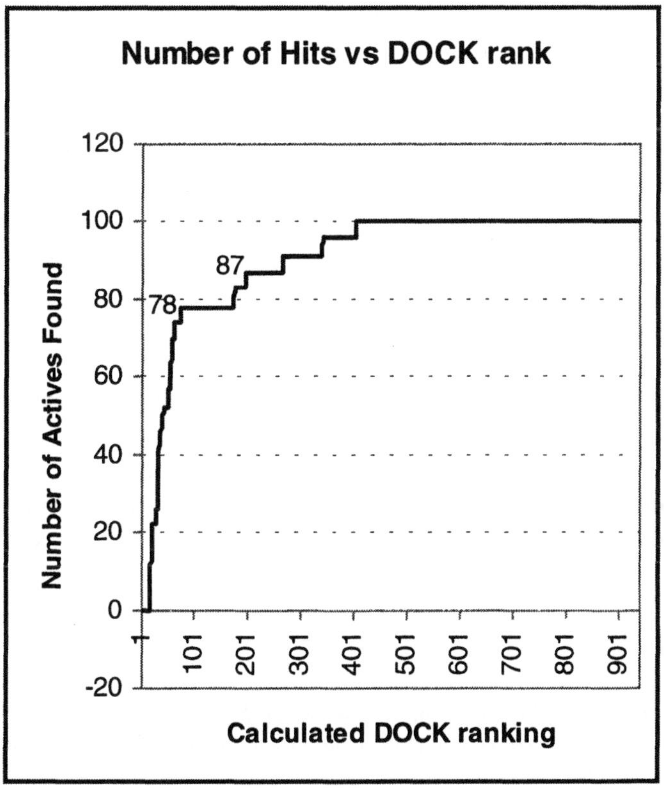

Figure 18. Number of Actives Found vs DOCK rank. As highlighted in this graph, DOCK found 78% of the compounds that were experimentally active at 330nM in its 75 top ranked compounds (out of 1000), and found 85% of the known actives in its top 200 ranked compound

The overall results show that the combinatorial DOCK can be useful in designing libraries, providing as much as an enrichment factor of 4. It should be noted that these results cannot be readily compared with those from combiBUILD, since combiBUILD chose this set of 1000 from an original pool of 1 billion possibilities, and there are no experimental results on the compounds not chosen.

A last approach for combinatorial libraries that is being developed is a combined DANTE/combiBUILD program. This program will use DANTE to design a pharmacophore model with a shrink wrap surface. The surface will be mapped onto a three-dimensional grid, which can be used in a combiBUILD search for ligands. A heuristic scoring scheme can be used which generates positive scores for 1) functional groups binding according to the pharmacophore model 2) exploring "terra incognita". Scores will be penalized for compounds entering into "forbidden regions". In this way, a structure-based search of the possible ligands can be used to simultaneously match the known SAR and guide diversity into "terra incognita".

6. CONCLUSION

This chapter has reviewed the basic principles of computer-aided drug design, and several strategies of how it can be successfully integrated with combinatorial chemistry to develop highly effective site-focused libraries. Diversity plays a key role, as the more diverse set of compounds tested that fit the site-focused criteria, the more information is retrieved to improve the site-focused definition, which further directs the search in diversity space. In addition, if good 'hits' are found, the information can be fed back to find compounds close in diverse space to the hit. This new paradigm for structure-based combinatorial chemistry should provide a powerful tool for rapid discovery of novel, potent lead compounds in the years to come.

REFERENCES

1. Greer, J., Erickson, J.W., Baldwin, J.J. and Varney, M.D. Application of the three-dimensional structures of protein target molecules in structure-based drug design. *J. Med. Chem.*, 1994, **37,** 1035-1054.
2. Lam, P.Y.S., et al. Rational design of potent, bioavailable, nonpeptide cyclic ureas as HIV protease inhibitors, *Science,* 1994, **263,** 380-384.
3. Bohacek, R.S., McMartin, C. and Guida, W.C. The art and practice of structure-based drug design. *Med. Res. Rev.*, 1996, **16,** 3-50.
4. Leach, A.R. (1996). *Molecular Modelling: Principles and Applications.* Longman Pub. Group, White Plains, NY.
5. Marrone, T.J., Briggs, J.M. and McCammon, J.A. Structure-based drug design: computational advances. *Annu. Rev. Pharmacol. Toxicol.,* 1997, **37,** 71-90.
6. Kuntz, I.D., Meng, E.C., and Shoichet, B.K. Structure-based molecular design. *Acc. Chem. Res.,*1994, **27,** 117-123.

7. Ring, C.S. *et al.* Structure-based inhibitor design by using protein models for the development of antiparasitic agents. *Proc. Natl. Acad. Sci.,* 1993, **90,** 3583-3587.
8. Cohen, N.C. Molecular modeling software and methods for medicinal chemistry. *J. Med. Chem.,* 1990, **33,** 883-94.
9. Gould, R.G.E. Biological Correlations - The Hansch Approach. *Advances in Chemistry,* 1972, **114,** The American Chemical Society.
10. Brown, R.D. and Martin, Y.C. Use of structure-activity data to compare structure-based clustering methods and descriptors for use in compound selection. *J. Chem. Inf. Comput. Sci.,* 1996, **36,** 572-584.
11. Brown, R.D. and Martin, Y.C. The information content of 2D and 3D structural descriptors relevant to ligand-receptor binding. *J. Chem. Inf. Comput. Sci.,* 1997, **37,** 1-9.
12. Van Drie, J.H., Weininger, D. and Martin, Y.C. ALADDIN: An integrated tool for computer-assisted molecular design and pharmacophore recognition from geometric, steric, and substructure searching of three-dimensional molecular structures. *J. Comp. Aided Mol. Design,* 1989, **3,** 225-251.
13. Loew, G.H.,Villar, H.O. and Alkorta, I. Strategies for indirect computer-aided drug design. *Pharmaceutical Research,* 1993, **10,** 475-486.
14. Richards, F.M. Areas, volumes, packing, and protein structure. *Ann. Rev. Biophys. Bioeng.,* 1977, **6,** 151-176.
15. Connolly, M.L. Solvent-Accessible Surfaces of Proteins and Nucleic Acids. *Science,* 1993, **221,** 709-713.
16. Kuntz, I.D., Blaney, J.M., Oatley, S.J., Langridge, R. and Ferrin, T.E. A geometric approach to macromolecule-ligand interactions. *J. Mol. Biol.,* 1982, **161,** 269-288.
17. Goodford, P.J. A computational procedure for determining energetically favorable binding sites on biologically important macromolecules. *J. Med. Chem.,* 1985, **28,** 819-857.
18. Miranker, M. and Karplus, M. Functionality maps of binding sites: a multiple copy simultaneous search method. *Proteins Struct. Fun. Genet.,* 1991, **11,** 29-34.
19. Danziger, D.J. and Dean, P.M. Automated site-directed drug design: a general algorithm for knowledge acquisition about hydrogen-bonding regions at protein surfaces. *Proc. Roy. Soc. London. Series B,* 1989, **236,** 101-113.
20. Bohm, H.J. LUDI: rule-based automatic design of new substituents for enzyme inhibitor leads. *J. Comput.-Aided Mol. Des.,* 1992, **6,** 593-606.
21. Bartlett, P.A., Shea, G.T., Telfer, S.J. and Waterman, S. In *Molecular Recognition in Chemical and Biological Problems.* Royal Chemical Society, 1989, London, pp. 182-196.
22. Lauri, G. and Bartlett, P.A. CAVEAT: a program to facilitate the design of organic molecules. *J. Comput.-Aided Mol. Des.,* 1994, **8,** 51-66.
23. Roe, D.C. and Kuntz, I.D. What is structure-based drug design? *Pharm. News,* 1995, **2,** 13-15.
24. Miller, M.D., Kearsley, S.K., Underwood, D.J. and Sheridan, R.P. Flog- a system to select quasi-flexible ligands complementary to a receptor of known three-dimensional structure. *J. Comput.-Aided Mol. Des.,* 1994, **8,** 153-174.
25. Leach, A.R. Ligand docking to proteins with discrete side-chain flexibility. *J. Mol. Bio.,* 1994, **235,** 345-356.
26. Judson, R.S. *et al.* Docking flexible molecules - a case study of three proteins. *J. Comp. Chem.,* 1995, **16,** 1405-1419.
27. Oshiro, C.M., Kuntz, I.D. and Dixon, J.S. Flexible ligand docking using a genetic algorithm. *J. Comput.-Aided Mol. Des.,* 1995, **9,** 113-30.

28. Lewis, R.A. *et al.* Automated site-directed drug design using molecular lattices. *J. Mol. Graphics,* 1992, **10,** 66-78.
29. Roe, D.C. and Kuntz, I.D. BUILDER v2: improving the chemistry of a de novo design strategy. *J. Comput.-Aided Mol. Des.,* 1995, **9,** 269-282.
30. Moon, J.B. and Howe, J.W. Computer design of bioactive molecules: a method for receptor-based de novo ligand design. *Proteins Struct. Fun. Genet.,* 1991, **11,** 314-328.
31. Bohacek, R.S. and McMartin, C. Multiple highly diverse structures complementary to enzyme binding sites - results of extensive appplications of a de novo design method incorporating combinatorial growth. *J. Am. Chem. Soc.,* 1994, **116,** 5560-5571.
32. Rotstein, S.H. and Murcko, M.A. GroupBuild: a fragment-based method for de novo drug design. *J. Med. Chem.,* 1993, **36,** 1700-1710.
33. Bohm, H.J. The computer program LUDI: a new method for the de novo design of enzyme inhibitors. *J. Comput.-Aided Mol. Des.,* 1992, **6,** 61-78.
34. Eisen, M.B., Wiley, D.C., Karplus, M. and Hubbard, R.E. HOOK: a program for finding novel molecular architectures that satisfy the chemical and steric requirements of a macromolecule binding site. *Protein Struct. Funct. Genet.,* 1994, **19,** 199-221.
35. Bohm, H.J. Computational tools for structure-based ligand design. *Prog. Biophys. Molec. Biol.,* 1996, **66,** 197-210.
36. Kollman, P.A. Free energy calculations: Application to chemical and biochemical phenomena. *Chem. Rev.,* 1993, **93,** 2395-2417.
37. Still, W., Tempczyk, A., Hawley, R.C. and Hendrickson, T. Semianalytical treatment of solvation for molecular mechanics and dynamics. *J. Am. Chem. Soc.,* 1990, **112,** 6127-6129.
38. Wesson, L. and Eisenberg, D. Atomic solvation parameters applied to molecular dynamics of proteins in solution. *Protein Science,* 1992, **1,** 227-235.
39. Holloway, M.K. *et al.* A priori prediction of activity for HIV-1 protease inhibitors employing energy minimization in the active site. *J. Med. Chem.,* 1995, **38,** 305-317.
40. Bohm, H.J. The development of a simple empirical scoring function to estimate the binding constant for a protein-ligand complex of known three-dimensional structure. *J. Comput.-Aided Mol. Des.,* 1994, **8,** 243-256.
41. Wipf, P., Cunningham, A., Rice, R.L. and Lazo, J.S. Combinatorial synthesis and biological evaluation of library of small-molecule ser/thr-protein phosphatase inhibitors. *BioOrg. Med. Chem.,* 1997, **5,** 165-177.
42. Graybil, T.L. *et al.* In *Molecular diversity and combinatorial chemistry: libraries for drug discovery.*, ACS Monograph, 1996, American Chemical Society, Washington, pp. 16-27.
43. Sheridan, R.P. and Keasley, S.K. Using a genetic algorithm to suggest combinatorial libraries. *J. Chem. Inf. Comput. Sci.,* 1995, **3,** 310-320.
44. Kick, E.K., Roe D.C. *et al.* Structure-based design and combinatorial chemistry yield low nanomolar inhibitors of cathepsin D. *Chemistry and Biology,* 1997, **4,** 297-307.
45. Murray, C.W. *et al.* PRO_SELECT: combining structure-based drug design and combinatorial chemistry for rapid lead discovery. 1. Technology. *J. Comput.-Aided Mol. Des.,* 1997, **11,** 193-207.
46. Van Drie, J.H. An inequality for 3D database searching and its use in evaluating the treatment of conformational flexibility. *J. Comput.-Aided Mol. Des.*, 1996, **10**, 623-630.
47. Van Drie, J.H. Strategies for the determination of pharmacophoric 3D database queries. *J. Comput.-Aided Mol. Des.*, 1997, **11**, 39-52.
48. Van Drie, J.H. "Shrink-Wrap" Surfaces: A New Method for Incorporating Shape into Pharmacophoric 3D Database Searching. *J. Chem. Inf. Comput. Sci.*, 1997, **37**, 38-42.

49. Li, J., Murray, C.W., Waszkowycz, B. and Young, S.C. Targeted Molecular diversity in drug discovery: integration of structure-based design and combinatorial chemistry. *Drug Discovery Today,* **1998, 3,** 105-112.
50. Available Chemistry Database, version 93.2. Distributed by Molecular Design Limited, San Leandro, CA.
51. Weiner, S.J. *et al.* A new force field for molecular mechanical simulation of nucleic acids and proteins. *J. Am. Chem. Soc.,* 1984, **106,** 765-784.
52. Sun, Y., Ewing, T.J.A., Skillman, A.G. and Kuntz, I.D. Combinatorial DOCK: Structure-based Combinatorial Docking and Library Design, *J. Comput.-Aided Mol. Des.*, 1998, in press.
53. Skillman, A.G., *Personal Communication.*

Chapter 8

Managing Combinatorial Chemistry Information

Managing Information

Keith Davies[1] and Catherine White[2]

1. Research Fellow, Department of Chemistry, University of Oxford, UK

2. Oxford Molecular Group, Oxford, UK

Key words: Database, Library, ORACLE, Pharmacophore

Abstract: An architecture for managing Combinatorial Chemistry information is presented and discussed in the context of selected applications. This recognises the need to improve the chemists' productivity. Issues relating to the increase in the numbers of reagents and products used are explored: the conclusion is that it is necessary to take a more formal approach to design and integration of chemical and biological information management systems.

1. INTRODUCTION

Pharmaceutical and Agrochemical research have seen a dramatic improvement in productivity in the last 3-4 years. The number of compounds which are made and tested has increased dramatically with the adoption of Combinatorial Chemistry for synthesis and High Throughput Screening for biological testing. In addition, driven by the goal of reducing cost, technology has been adopted which allows reduced quantities to be tested and less thorough practices used to confirm the purity and chemical structure. This evolution has demanded a lowering of standards in discovery from using single pure samples in highly reproducible experiments. This has been difficult for many scientists but is trivial compared to the metamorphosis in working practices required to achieve the next significant improvement in productivity.

Traditionally, Chemical Information Systems have been used to archive structures and associated information after completion of the synthesis.

Some use has also been made of reaction databases but mainly as a means of locating relevant synthetic publications. Thus the synthetic chemist is accustomed to using computer software such as ISIS [1] to record a subset of the information which has previously been entered into a hand-written notebook. The electronic information is almost invariably too little to repeat the synthesis, to confirm the chemical structure or help with making other analogues in a project. Most companies do not record any data in a shared database prior to the chemical structure, purity and other data such as quantity meeting certain standards. It is therefore not surprising that, many chemists view using such software as a chore which reduces their productivity. The idea that a computer program might help a chemist do his or her job is quite a revolution and requires a complete change of working practices.

Before continuing, it is necessary to briefly review the chemist's job. One may naively assume that a chemist is employed to make molecules with the desired biological activity. While this feeds his or her *ego*, it is neither the most certain nor the most rapid route for promotion. In the 1980s, most major pharmaceutical companies evaluated the effectiveness of their chemistry departments by how many molecules they made per chemist per year. Even prior to the adoption of Combinatorial Chemistry some adjustments were made for the effort required and this approach continues to be widely used. In defence of management, it must be realised that chemists sometimes move from project to project and the time-scales for discovery are often significantly longer than those appropriate to review individual and group performance.

The immediate driving force for a quantum improvement in Chemical Information System is therefore to cope with between 10 and 1000 times more compounds being made per year by each chemist. In addition, there are information management issues concerning molecules which were considered for synthesis but excluded on consideration of properties, diversity, synthetic accessibility, price etc. Recently, there has been increasing consideration of reducing time-scales by being cleverer – so called Smart HTS and Smart Combinatorial Chemistry. Companies using Smart HTS, consider testing or making only an appropriate subset of the conceivable compounds and review the order in which these are tested. The same basic principle of being selective can also be applied to synthesis when only the products of some combinations of reagents are made using carefully programmed robots to generate what may be referred to as a sparse library.

1.1 Data, Information and Knowledge

In principle, there are a number of ways of classifying, grouping or relating the various parameters, ideas, properties or results which may be stored in a database. An obvious distinction is between calculated and measured properties. However, the method used to generate the property is strictly speaking an attribute of that property and even for measured properties it is not usually the raw experimental data that is stored. These aspects are important to consider when designing database systems and when making decisions based on the stored data. A further useful concept is the distinction between data, information and knowledge. In general, it is knowledge that is required to make a decision and this is deduced from the information which in turn was derived from some data. This is represented schematically in figure 1.

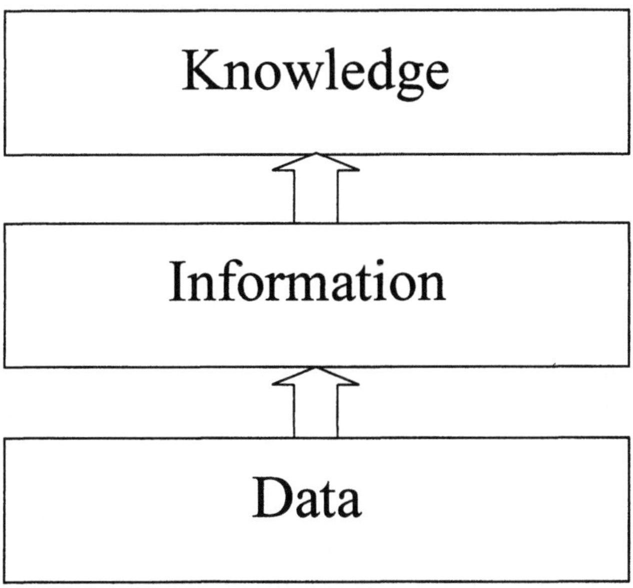

Figure 1. The relationship between Data, Information and Knowledge

The information derived from the data normally uses less space to be stored, is more meaningful but is insufficient to regenerate the original data. It can be useful to define information as a *transformation* of the data from which original data may not be regenerated. In a similar manner, defining knowledge as something deduced using logic from information (or data) is valuable when designing database systems. For example, an IR spectrum is data, but the presence of peaks at given frequencies provides the some information about which functional groups may be present. These in turn may allow the knowledge that a re-arrangement occurred during a reaction to be deduced.

In modern approaches to drug discovery there are numerous approximations, assumptions and accuracy or error limitations. Inevitably, this affects the reliability of the data, information and knowledge with the obvious implication that some of the knowledge will be incorrect. Appreciating the differences between the types of parameters can help to ensure that when possible the appropriate statistical techniques are used and that the reliability of knowledge is considered when making important decisions.

1.2 Strategic Goals

For a Chemical Information System to be widely used and highly valued, it should *help each chemist do his or her job.* In the next century, one may envisage a system which would answer the ultimate question: which molecule should I make to exhibit the desired activity with minimum undesirable side-effects. Today, a realistic goal is to have a computer system which increases the number of molecules the chemist can justify making and reduces the time taken.

In the 1980's a significant benefit of Chemical Information Systems was to avoid making the same molecule more than once inadvertently. However, when using Combinatorial Chemistry avoiding attempting to make individual molecules may not be cost-effective: the cost of moving products around on testing plates or programming the synthesis robots may be prohibitive. Obviously, inadvertently making an identical library to one previously made needs to be avoided and where there is partial overlap between a proposed library and one already made there may be a need to reconsider the choice of reagents. It can therefore be asserted that an important goal is to be able to compare the contents of libraries.

However, this assertion is based on the assumption that the compounds intended to be made were actually made in sufficient quantities and to adequate levels of purity. The yields in many Combinatorial Chemistry synthesis vary significantly and the purity may also be problematic. The additional steps required to purify and measure the quantity generated are not usually necessary to establish whether the major product shows any significant biological activity. It is therefore not surprising that many companies perform quality control checks on a subset of their samples. This prohibits accurate decisions on the overlap of libraries.

It may be appropriate to select either reagents or products by calculation or measured properties. It is therefore often appropriate to store the properties which are frequently used if the Chemical Information System is going to help chemists do their job more productively. Calculated properties which are only used occasionally may not justify storage if they can be quickly generated when required. This approach implicitly requires storing information for reagents which have never been used and products which were never made – which is quite a change from traditional thinking! It should be noted that adopting this approach does not require all data or all information to be stored. This will be illustrated later when the connection tables of the products are not permanently stored although properties of the corresponding molecules are stored.

Analysis of the time spent planning, rehearsing and making libraries can identify scope to improve productivity and also what information may be of most value to other scientists. Ordering and managing the inventory of reagents can be time-consuming and often requires additional functionality (discussed later) not found in older inventory databases.

Often the time taken to rehearse the reaction conditions etc to obtain optimum yields for range of diverse reagents is very significant. Thus recording and storing the yields for a variety of reagents and reaction conditions is valuable information to share with colleagues. Exchanging this type of data requires changes to the way chemists work and a readiness to accept compact ways of summarising information.

2. ARCHITECTURE

2.1 Relational Databases

Traditionally data, properties, information etc has been stored in files on computer disks. More recently, it has become common practice on Macintosh computers, when using Microsoft software or some UNIX applications, to use either extensions to the file name or the first few bytes in the file (or another file) to indicate some aspects of the data, for example that it is suitable for Microsoft Excel. While this approach is practical to indicate something about files containing columns of data, it is not appropriate to store information about the values in cells in spreadsheet or how it relates to data in other columns. This requires a relational database such as ORACLE, and for performance reasons the values in the cells may only be accessed via the ORACLE API (Application Programming Interface) or SQL (Standard Query Language). In other words, it is suggested that relational databases such as ORACLE should be viewed as sophisticated file systems which allow the values to be organised, efficiently stored, rapidly retrieved etc.

The use of ORACLE to store the atomic data comprising a molecule does not necessarily mean that ORACLE has to be used to search it. In some cases, such as 3D searching the ORACLE standard text and numeric search engines offer nothing of use and proprietary applications have to be used. It is important to realise that the term database refers to a convenient subset of values or ORACLE tables not to specific files.

The architecture used by the Chem-X [2] system for Combinatorial Chemistry can be summarized in figure 2.

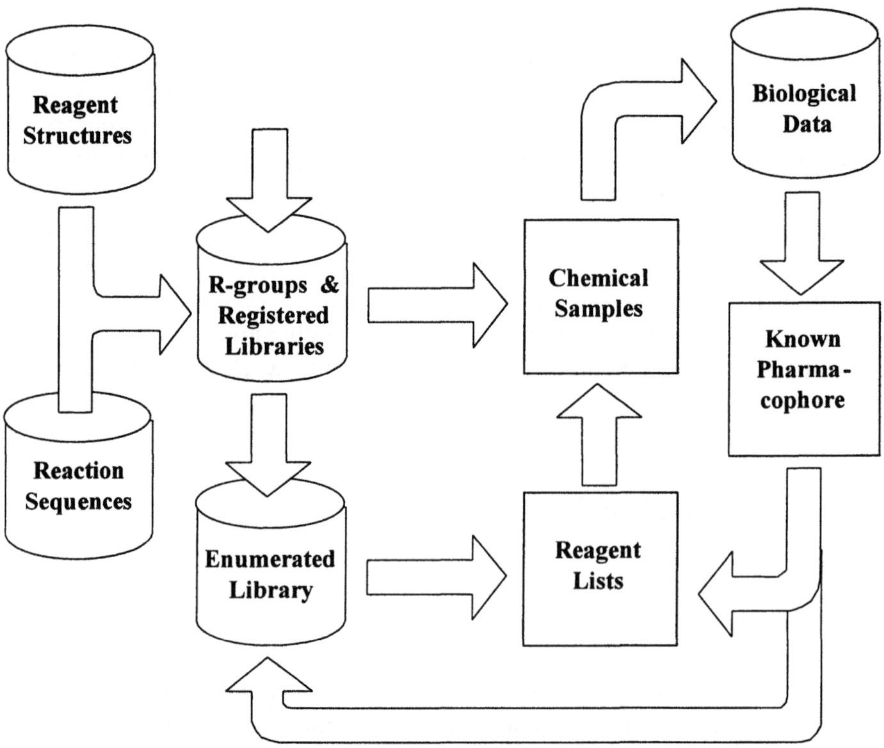

Figure 2. Combinatorial Chemistry Information Architecture

The relationships between the various data which may be stored are important to avoid unnecessary redundancy and deliver optimum performance. The reagents used in the synthesis of a library may have some functional groups in common but are not necessarily otherwise similar. For a given library, the chemical reactions are the same but sub-libraries may have differing reaction conditions. In other words, it is the common chemical reactions which group a set of products into a combinatorial library.

This definition is important because it means that an unexpected product such as the result of a rearrangement is treated as an exception in the same way as products arising from impurities etc. From a chemical information perspective, this means that a library is defined by one or more transformations from reagents to R-groups and the way in which the R-groups are joined together.

2.2 Reaction Databases

When using Combinatorial Chemistry on a large scale, it can be appropriate to think in terms of manufacturing a large number of samples to be tested using a significant degree of automation. Each synthesis requires a sequential series of reactions (some of which may be multiple step reactions). Recording the reaction conditions and product data such as yield, purity etc may help other chemists attempting to generate libraries in the future.

Traditionally, reaction databases have stored explicit data showing how molecules react to give a new product storing all atoms for the molecules. For Combinatorial Chemistry, it is important to store generic structures of the type shown in figure 3 which represent the transformation that the reaction achieves. Furthermore, using the approach adopted by Chem-X, this information is automatically used to search for and select suitable reagents and generate the R-groups to be stored in the Library Database.

For most real libraries, there are alternative generic reactions which may differ in the choice of final product (which forms the "core" of the library). In addition, there may be other constraints relating to the choice of reagent necessary to obtain adequate yields or purity, for example, the maximum allowed steric crowding or the desired degree of electrophilicity. At present, although the architecture shown could store this type of knowledge, the version of Chem-X available does not store or use this knowledge.

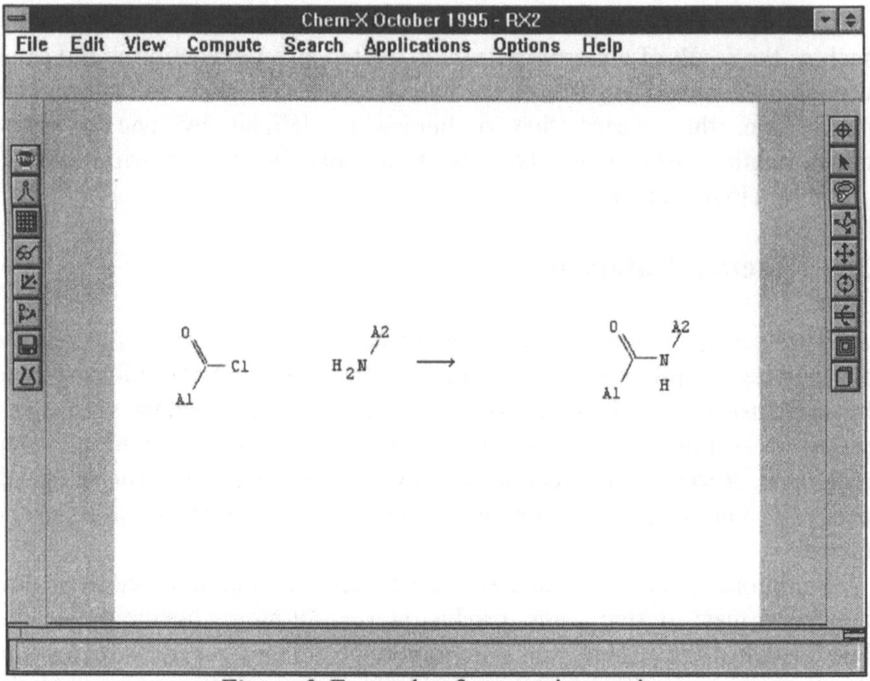

Figure 3. Example of a generic reaction

For each reaction, it is appropriate to store the libraries for which it is used, the step number in the sequence of reactions for that library and appropriate synthetic information. This may consist of a reference to a laboratory notebook or computer file but summary conditions may be more helpful to other chemists within an organisation.

Table 5. Typical reaction database information

Column	Value
Reaction Identifier	RX1
Library Identifier	EKD01
Reaction Sequence Number	2
R-group Number	3
Quantity	0.01
Solvent	DMSO
Time	5

2.3 Reagent Databases

Many chemists have been surprized to discover that their reagent database applications have not proved adequate for Combinatorial Chemistry. In some instances, the required search queries can not be answered. For example, "list the 10 most similar reagents to a given compound which can be delivered next week within budget and of the desired purity". Other companies have experienced difficulties because of the lack of automation or inadequate performance of the software. Much can be gained from the experiences of the manufacturing industry and the working practices and systems they require to ensure smooth continuous production. Thus the essential information to be stored for each reagent is

- Supplier
- Price
- Quality eg Purity or Grade
- Quantity
- Delivery Timescale

Table 2 shows a typical set of tables used by Chem-X.

Table 6. Typical Relational Database Columns used by Chem-X

Field name	Type	Description
STR_FULLNAME	Text	Chemical structure full IUPAC name
STR_TRIVIAL	Text	Chemical structure trivial name
STR_FORMULA	Text-formula	Molecular formula
STR_WEIGHT	Text-numeric	Molecular weight
STR_COMMENT	Text	Free text comment
STR_PROJECT	Text	Keyword for originating project
STR_HAZARD	Text	Keyword for hazard
STR_COSHH	Text	COSHH reference code
STR_ACTIVE	Text-numeric	Percentage active fragment (<100% for salts)
STR_CAS	Text	CAS number
STR_QUANTITY	Text-numeric	Total quantity
STR_MINLEVEL	Text-numeric	Minimum stock level for automatic re-ordering
STR_LOCATION	Text	Storage location when in stock room
STR_TYPE	Text	Class of compound (Stock or In-house)
STR_LOGP	Real	Reserved for LogP value
LOT_ID	Text	Batch identifier for individual bottle
LOT_ORDER_ID	Text	Order identifier
LOT_CHEMIST	Text	Chemist identifier (eg name)
LOT_DATEMADE	Text-date	Date batch registered
LOT_COLOUR	Text	Colour keyword (In-house only)
LOT_PSTATE	Text	Physical state keyword (In-house only)
LOT_QUANTITY	Text-numeric	Quantity in batch
LOT_UNITS	Text	Units for quantity
LOT_LOCATION	Text	Location keyword
LOT_MP	Text-numeric	Melting point (In-house only)
LOT_BP	Text-numeric	Boiling point (In-house only)
LOT_SOLVENT	Text	Solvent (In-house only)
LOT_SUPPLIER	Text	Supplier identifier (Stock only)
LOT_CATALOG	Text	Catalogue number (Stock only)
LOT_TRAY_ID	Text	Reserved for reagent tray identifier
LOT_ANALYSIS	Text	Analysis comments

LOT_CONCENT	Text	Reserved for component concentration of solution
LOT_COST	Text-numeric	Cost of batch (Stock only)
LOT_REQ_DATE	Text-date	Required delivery date (Stock only)
LOT_ORD_DATE	Text-date	Order date (Stock only)
LOT_USE_DATE	Text-date	Issue date (Stock only)
LOT_EXP_DATE	Text-date	Expiration date (Stock only)
LOT_PROJECT	Text	Keyword for originating project
LOT_HPLC	Text	HPLC analysis status
LOT_IRSPEC	Text	IR spectrum status
LOT_UVSPEC	Text	UV spectrum status
LOT_MSSPEC	Text	Mass Spectrum status
LOT_1HNMR	Text	Proton NMR status
LOT_13CNMR	Text	13C NMR status
LOT_CHN	Text	Carbon-Hydrogen-Nitrogen atomic analysis status
LOT_EXAMASS	Text	Experimental exact molecular mass
LOT_H2OSOL	Text-numeric	Solubility in water (optional)
LOT_DMSOSOL	Text-numeric	Solubility in DMSO (optional)

Some suppliers have crippled integration with Combinatorial Chemistry Information Systems, by changing unique identifiers for compounds over time, or by using different identifiers for the same compound, or removing compounds etc. Consequently, the only practical and safe approach is to assign a unique reagent identifier for each reagent used, regardless of supplier, and for each source store the alternative suppliers (including for reagents made in-house). In Chem-X, this same reagent identifier is also used for the R-groups in the Library Database.

In many companies, the adoption of Combinatorial Chemistry greatly increases the numbers of chemicals stored and the number of bottles for each chemical. In the future, it may be necessary to reserve bottles (or parts thereof) for the synthesis of a given library. Otherwise, it is conceivable that a library synthesis is delayed by the delivery of a specific reagent and by the time that reagent is delivered, another reagent is out of stock.

The inventory application of Chem-X, included options to automatically order chemicals as well as the more usual tracking functionality. Significant efficiency savings can be made using such systems as well as ensuring the appropriate cost controls. Some companies favour the storage of standard solutions of selected sets of commonly used reagents such as amines. An inventory of the location and quantity of these sets is also possible using Chem-X.

2.4 Library Databases

In order to share information about which compounds and libraries were not made and why, it is essential to register libraries prior to attempting synthesis. The information recorded at this time must include at least the reactions and reagents used. This also facilitates the automation of the synthesis. Some companies favour the storage of virtual libraries, referring to reagents and products which have not yet been made, in separate databases from real libraries which have been made. This has an unfortunate consequence that every library that is made has to be registered twice and so do the corresponding reagents! It also encourages a working practice which prohibits sharing ideas and failures which are not usually in the company's interests. Repeating somebody else's failed experiences can be very time-consuming and expensive. The alternative is to store the knowledge whether a reagent or library is real as a status in much the same way as whether a quality check indicated whether the synthesis of a given product was successful. This does require a reagent database such as an inventory application to have the ability to store compounds which have not been ordered or delivered. It also means that plates or vials containing samples which have been successfully synthesised have to be registered.

In Chem-X, the Library Database contains relational tables recording the use of each R-group in a library in terms of the R-group number and position in the list of alternatives for each library. During library registration a row of data is inserted for each component. An enumerated database containing the connection tables of the products may also be generated at this time. Enumerated databases are required for some library design calculations such as those using pharmacophore diversity concepts.

Table 7. Typical information store in a Chem-X Library Database

Field name	**Description**
CC_LIB	Library identifier
CC_RGP_NO	R-group Number
CC_RGP_ALT	Position in list of alternatives for R-group
CC_DATE	Date library registered
CC_FORMULA	Molecular formula for R-group
CC_WEIGHT	Molecular weight for R-group
CC_COMMENT	Comment field

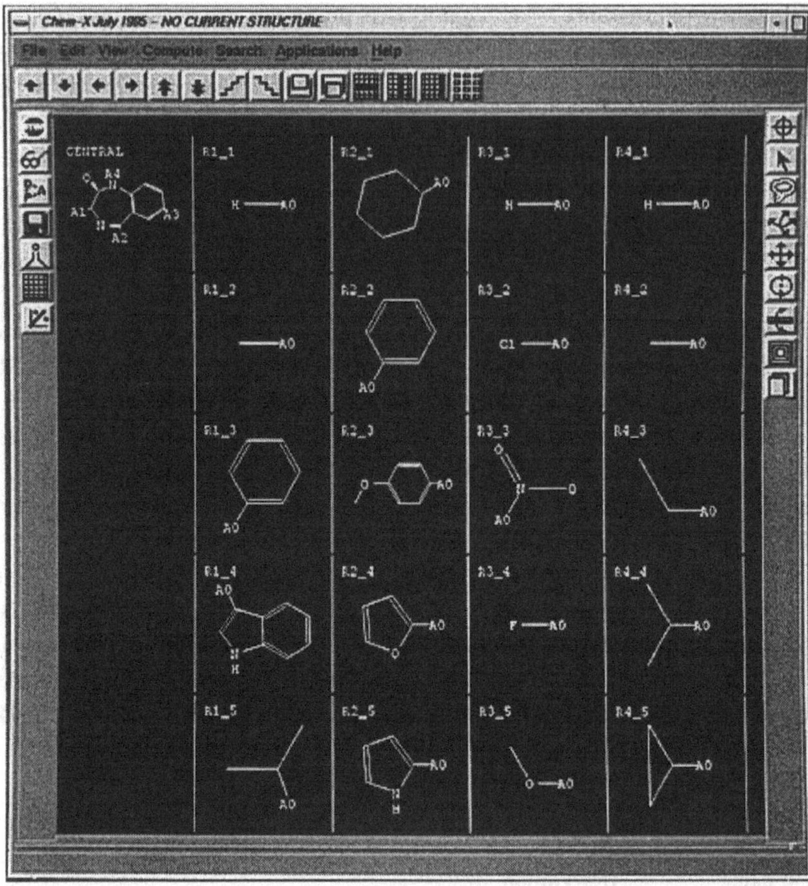

Figure 4. The components for a library.

Obviously, it is desirable to decide which R-groups to include prior to optimising the synthetic conditions. The integration with the Reagent Database and Inventory application may generate a subset of those available within cost or delivery constraints. Ultimately, most chemists wish to review the list. Chem-X includes the capability for the components or R-groups in a library to be viewed in a special layout where the components are arranged in columns as shown in figure 4. A new sublibrary may be generated by selecting, using the mouse, those R-groups to be retained or rejected. This approach is well suited to very large libraries especially when the products could not be displayed on a few pages.

2.5 Enumerated Databases

Enumerated databases contain the connection tables, 2D drawing coordinates and usually the 3D coordinates for the products. This information can be generated at any time and consequently, there is no requirement to store it permanently. Some companies do export these structures via SD files into other database applications such as ISIS or RS3 [3]. This allows faster searching but necessitates caution because the structures generated may not be present in the corresponding samples unless confirmed by analysis. It is also inevitable that for large libraries vast quantities of disk space are required.

Table 8. Typical information stored in an enumerated database

Field name	**Description**
CC_WEIGHT	Molecular weight of product
CC_FORMULA	Molecular formula of product
CC_SYSNAME	Concatenated alternatives (CC_RGRP_ALT) of R-groups
CC_NAME	Concatenated reagent IDs of R-groups
CC_COMMENT	Comment field

The main use of enumerated libraries is to calculate properties which are useful when deciding whether to make an entire library, use a given reagent or selectively make certain products. These calculations are required prior to synthesis and are often performed by computational chemists as a service to the organic chemists performing the synthesis. This information can also be used to generate files for custom interfaces to automatic synthesisers.

2.6 Searching

Traditional 2D database systems of samples for testing are commonly searched by substructure to select molecules to test, to avoid duplicate synthesis and to retrieve results for a series of structures (defined by substructure, activity, research project etc). The potential benefits of 3D databases has been limited by the costs of locating and re-plating physical samples for structures which match the search query or pharmacophore. For Combinatorial Chemistry this limitation can only be overcome by very large scale robotics systems. One might imagine generating a focused selection which meet some criteria such as have the ability to interact at a given receptor site. In such circumstances it is conceivable that it is cheaper and/or faster to repeat a synthesis for a subset of a library rather than locate an old sample (when one may discover that there is too little or it is not sufficiently pure for a secondary screening assay).

In Chem-X, the traditional 2D searches may be performed on the Library Database, when the connection tables of probable hits are generated on the fly and stored in an enumerated database. Alternatively, enumerated databases may be used when the searches are performed faster but time will have been used to enumerate the library. Typically, for 3D searches and library design it is usually preferable to enumerate the library.

Searches on the text and numeric values are performed by the ORACLE search engine using the Chem-X user-interface. This allows the user transparent access to ORACLE as well as to search substructure and text/numeric values at the same time. The details of how this is implemented are outside the scope of this publication.

2.7 Integration with Robotics

For small scale Combinatorial Chemistry there is no obvious reason to use robotics. A series of flasks and a shaker is normally sufficient for performing 6-20 reactions at one time. For larger numbers of reactions, when an inert atmosphere is necessary or when temperature control is required (heating or cooling) it is often preferable to use an automatic synthesizer such as the Advanced ChemTech 496. When products from all combinations of reagents are required, programming such systems is relatively straight forward but when only products from some combinations of reagents are required (to form a sparse library) the set up of such robotics systems can be tiresome and time consuming. Chem-X offers an interface to the ACT496 (and other similar robotics systems) which is especially useful when making large or sparse libraries. The interface computes the quantities of reagents required, allocates them to position in the rack of reagents and passes the dispense instructions for reagents into the reaction vessels to the robotics software. This saves significant time and reduces the scope for error.

2.8 Plates

Once the synthesis is complete, the products may be weighed and transferred to plates or vials for testing and storage. For most companies this requires a significant degree of manual sample handling and scope for error. Use of an electronic balance integrated with the information management system can ensure that procedures are followed, that values are only entered once and that appropriate bar codes are generated for any vials.

The ChemHTS-1 module of Chem-X provides integration to plate samples, copy plates, load analysis validation data and request biological testing. A simple set of ORACLE tables ensure speed and flexibility (tables 5-7).

Table 9. Tables used by ChemHTS-1: Structure Data Table

Field Name	Length	Comment
Library ID	16	Unique ID for each library (including virtual libraries)
Structure ID	36	Unique ID for each member within a library eg 1-2-3-4-5
Sample ID	8	Unique ID for each sample
Plate ID	8	Master plate ID unique for each plate
Well ID	3	Name of Well (A1...H12)
Mol Weight	8	Molecular Weight of expected product (row repeats for mixtures)
PercentageYield	4	Percentage yield
Weight	8	Weight (mg)
Volume	8	Volume (ml)
Status	8	Keyword for Chemical Sample Status

Table 10. Tables used by ChemHTS-1: Plate Data Table

Field name	Length	Comment
Plate ID	8	Plate ID
Creator ID	8	ID of Chemist, Analyst or Biologist who created the plate
Location	12	Normal storage location for plate
Creation Date	-	Date plate created
Notebook ID	8	Chemist/Biologist notebook reference
Parent ID	8	Plate ID of parent plate (if exists)
Project ID	8	Project for which plate was created
Plate Type	8	Type of plate eg MASTER or DILUTION
Plate Format	8	Name of plate format eg 80WELL
Mixture type	8	Keyword for mixture eg NONE, AXY, ABX etc
Priority	8	Text string for analysis or assay result
Required Date	-	Date analysis or assay result required
Status	8	Keyword Task Status

Table 11. Tables used by ChemHTS-1: Biological Data Table

Field name	Length	Comment
Sample ID	8	Sample ID
Plate ID	8	Plate ID for assay plate
Well ID	3	Name of well (A1 .. H12)
Assay ID	8	Assay ID
Result Type	8	Keyword eg SCALE3, IC50
Result	8	Numeric result
Concentration	8	Numeric concentration
Status	8	Keyword for biological result status

2.9 Mixtures

Some mixtures are the inevitable consequences of the reagents while others may result from the synthetic protocol. For instance, if some reagents contain the same reactive functional group more than once then it may react at more than one site giving alternative products. It is becoming common practice to avoid or protect such reagents when single compound samples are required. Early Combinatorial Chemistry synthesis always generated mixtures often using the split and mix approach which required little or no sophisticated robotics for large libraries. Unfortunately, testing mixtures can cause confusion because the activity may arise from many weakly active members and it is not possible to identify which molecule is active in an arbitrary mixture without re-testing single compounds. So called orthogonal mixtures avoid this issue by using the same molecule more than once. Discussion of the various experimental approaches are outside the scope of this chapter.

The storage architecture used by Chem-X allows mixtures to be stored simply by repeating the same plate and well identifier with a different structure identifier. This approach means that any mixtures can be stored regardless of how the mixture was generated. The software is configured such that when searches are performed all members of a mixture are searched and (if appropriate) included in the results.

2.10 Assays

The advent of HTS has meant that molecules are often tested in many more assays than was previously the case. This has the consequence that the explosion on biological results is frequently larger than that of chemical information. The discussion of biological data handling is outside the scope of this chapter but to have an effective drug discovery information system it is important that chemical and biological information are fully integrated. Again, for relatively small amounts of biological values and chemical information this is not a significant problem but for hundreds of thousands of chemicals and hundreds of assays many of the older systems have built in size limits or performance limitations which prohibit proper exploitation of the experimental results.

3. APPLICATIONS

3.1 Similarity

It is now universally acceptable to measure the similarity of two molecules using the Tanimoto Index of their database keys or fingerprints. There is no standard fingerprint and the similarity measures consequently can vary. Most fingerprints are dependent upon which functional groups are present and the connection path between the groups. The similarities are closest to those expected by medicinal chemists when the emphasis is on presence or absence of substructures defining functional groups and substitution patterns. Consequently one would expect that the database keys based on functional groups as used by the MACCS and Chem-X database systems are closer than those that are purely mathematically derived such as that used by the Daylight software [4]. The Chem-X software includes the ability to change the weighting on the various components of the fingerprint and change the functional groups used when computing the Tanimoto index thus allowing users to have the best of all worlds.

The use of fingerprints which indicate the presence or absence of functional groups can be very beneficial when eliminating reactive molecules or identifying required functional groups in active molecules. Consequently, it is desirable to store this information in a database in such a way as it can readily be viewed in a spreadsheet. It is not usually feasible to store the similarity of all pairs of molecules which have been synthesised yet alone all those that have been thought about. Good spreadsheet software should be able to compute and display such information on-the-fly.

3.2 Calculated Properties

Computational chemistry programs can calculate a range of molecular parameters with varying accuracy. Some properties such as LogP are appropriate to calculate only for the molecules being tested while other parameters such as molecular weight may be useful for the component reagents.

Properties can be used in 3 ways to select molecules to test:

- Minimum and maximum values which define a range outside which molecules are not appropriate for testing. Molecular Weight, LogP and solubility are commonly used in this way.
- Regular sampling to generate a selection which span a range for a given property for a molecule or a R-group such as volume.
- Similarity to the value for an active molecule as may be the case with dipole moment.

How properties are used and which properties may be appropriate depends on the nature and stage of the research programme. It is common practice to use a range when selecting molecules to test from a collection and regular sampling is only used when it is appropriate to test a subset of those molecules which are available. Obviously, similarity is only appropriate when some active molecules are already known and there is a hypothesis which defines which properties are important.

Many of the properties can be very quickly calculated, however, if the values are to be processed or viewed using software other than that which estimated the value, then it is necessary to store the values at least temporarily. For properties which are not dependent upon the conformation of the molecules this is technically feasible within most ORACLE environments but it is not usually feasible to store properties which change with the conformation as discussed later.

In addition to the value estimated, it is usually good practice to store the method which was used to generate it and an estimate of the possible error. Without this information, it is very easy to draw incorrect conclusions when reviewing tables of values with associated activity information. The shortage of statistical methodology to review this data including coping with estimated errors in the activity information is outside the scope of this chapter. A typical third normal representation table is shown in table 8. When the same property is stored for an entire database, the use of separate columns for each property may be preferred. However, while this may be appropriate for all molecules within a project, it is rarely the case for many properties within the entire collection. One disadvantage of the approach shown in table 8 is that a view may be required in order to manipulate the data in some spreadsheets.

Table 8. Typical table for storing properties

Field Name	**Description**
Library ID	Identifier for the library
Structure ID	Identifier for structure
Property	Calculated property value
Method	Method used for property calculation (identifier or keyword)
Error	Estimated error
Date	Date of calculation
Status	Keyword for property value status

3.3 Conformational Dependence

Computer scientists have been inclined to suggest that each conformation should be stored with its X,Y,Z coordinates, like different molecules. This demonstrates the frequent misunderstanding that many drug-like molecules are extremely flexible and may have hundreds or thousands of low energy significantly different conformations, any one of which may be responsible for the exhibited biological activity. Furthermore, some properties are conformational averaged properties whereas others are needed for the conformation which interacts with the receptor.

Most computational software programs store sets of conformers in terms of their internal coordinates or torsion angles for the bonds which are being rotated. This reduces the space required to store the information very dramatically – typically in Chem-X from 2 Kbytes to 4 bytes per rotating bond. At present, as far as the author is aware, this type of information is not being stored and shared between chemists using ORACLE databases. Instead, proprietary databases are being used which can be quickly scanned to generate any conformationally dependent property which is then made available in ORACLE.

Some software, including Chem-X, has the ability to identify conformations which may fit within a receptor site and favourably interact. These conformations may also be usefully stored within a database for visualisation and comparison of calculated properties.

3.4 Pharmacophores

In recent years, Chem-X is probably best known for its pharmacophore approach to determining *molecular diversity*. This phrase can be quite misleading to a beginner because the diversity which is of interest is that for an entire set of molecules not the diversity of a given molecule which is sometimes better described as its *promiscuity*.

The pharmacophore approach is capable of quantifying the promiscuity for each molecule in terms of the list of the number, type and geometry of the pharmacophores exhibited and diversity of a library in terms of the sum list for all the molecules in the library.

The definition of the pharmacophore is important in order to obtain meaningful estimates of promiscuity and diversity. In Chem-X, the pharmacophore is defined in terms of the types of interaction with the receptor and the distance between the centres of these interactions. In the Jan98 release of Chem-X, 7 centre types were used:

- Hydrogen bond donor
- Hydrogen bond acceptor
- Positively charged centre
- Aromatic ring centre
- Lipophilic centre
- Acidic centre
- Basic centre

The standard configuration does not use a negatively charged centre because all such atoms are also hydrogen bond acceptors (whereas all positively charged centres are not hydrogen bond donors). The software allows pharmacophores to be defined by either 3 interaction centres (forming a triangle) or 4 interaction centres (forming a tetrahedron) with consequential 3 or 6 distances between these centres. For most flexible drug-like molecules there are many pharmacophores which may be exhibited by low energy conformations. Storing the pharmacophores compactly can be problematic and consequently a bit mask is favoured which also forms a pharmacophore fingerprint analogous to the functional group fingerprint discussed earlier. Each bit in the fingerprint corresponds to a specific selection of centre types and a set of distances between the centres. The distances used by default are up to 15 Angstroms and this is divided into 31 distance ranges each represented by different bits. In practice, the size for a 3 centre pharmacophore fingerprint is less than 7x7x7x31x31x31 because some of the sets of distances do not form a triangle (the longest side is longer than the sum of the shorter sides). The space used is further reduced by recognising the symmetry of the pharmacophores. If these fingerprints are stored without compression then a typical 3 centre pharmacophore fingerprint is approximately 800Kbytes whereas 4 centre pharmacophores may be up to 50Mbytes.

Obviously, storing all the possible 4 centre pharmacophores for collections of millions of molecules becomes impractical. For molecules which are not very promiscuous and consequently set few bits in the pharmacophore fingerprint, compression algorithms may substantially reduce the space required. If storing compressed 3-centre pharmacophore fingerprints required less than about 50Kbytes each, then it becomes feasible to save on disk the pharmacophores exhibited by up to a million molecules. For 4-centre pharmacophores, it is necessary to restrict the storage of pharmacophores to hundreds of thousands of molecules such as those associated with a given project or virtual library. Although it may be useful to archive and repeatedly access this information, companies which have explored storing such information have not used ORACLE to handle the data. It is probable that performance and configuration limitations would be encountered with at least some molecules (such as those requiring more than 64Kbytes to store the compressed fingerprint).

Visualising fingerprints in terms of the bit patterns does not usually lead to innovative thoughts. However, visualisation techniques have been developed for 3 centre keys which display in 3D a cube with each axis of the cube representing the distance along the side of the triangle forming the pharmacophore. Each point in the cube then represents a specific geometry of a triangle and symbols can be used to differentiate between the various possible combinations of 3 centres chosen from the possible 7. This approach may be used to visualise the promiscuity of a single molecule or the diversity of a set of molecules such as those forming a library as shown in figure 5.

In drug discovery, it can be useful to perform logical operations on the lists of pharmacophores exhibited by a series of active molecules. If all the molecules interact at the same receptor in the same way then an AND operation will reveal at least one pharmacophore which is common to all the molecules. Given sets of active molecules and inactive molecules, the pharmacophores which are exhibited by the active set but not the inactive set may be of interest. However, the reader is cautioned that inactivity may arise due to reasons other than failure to exhibit the desired pharmacophore: the biological data may contain experimental errors and some approximations may have been made in computational methods used. This may lead to erroneous conclusions.

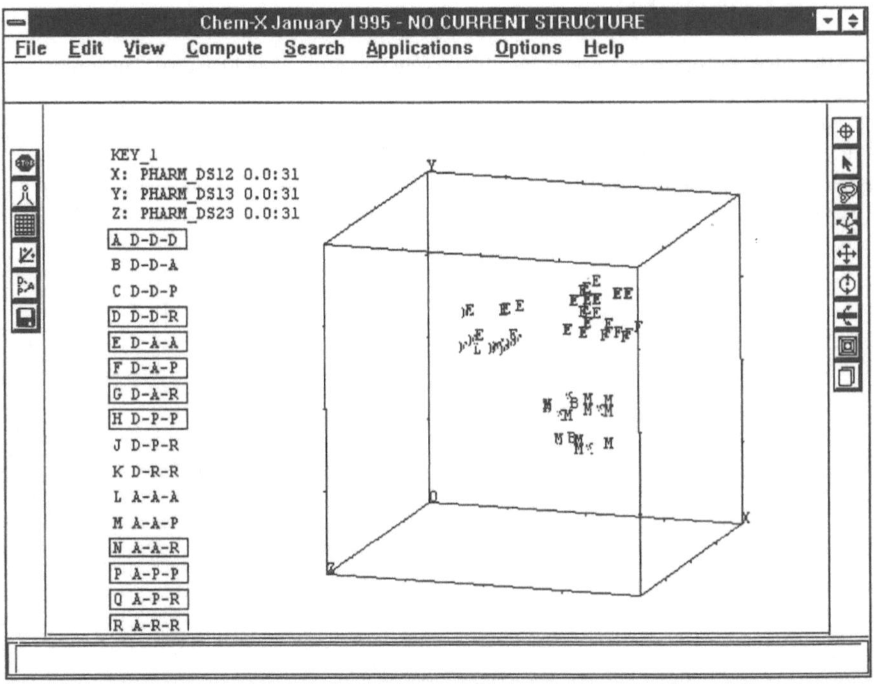

Figure 5. A pharmacophore plot

4. CONCLUSIONS

The architecture discussed has been based on the premise that the information generated is valuable because it will be used to make decisions about which molecules to make and test. This requires that the entire drug discovery process has an integrated information system which captures all the required data including that relating to molecules which were considered but not made. It is usually not possible to achieve this by "gluing together" existing old applications which were designed to handle orders of magnitude fewer records. In addition to quality information systems, adopting a *Smart* approach where intelligent decisions to make a subset of molecules of those which could be made, requires a significant change in the approach which has been traditionally pursued by medicinal chemists. Furthermore a level of computer literacy and theoretical chemistry not usually associated with organic chemists is a pre-requisite.

REFERENCES

[1] ISIS, MDL Information Systems Inc, 14600 Catalina Street, San Leandro, California 94577

[2] Chem-X, Chemical Design, Medewar Centre, Oxford Science Park, Oxford

[3] RS3, Oxford Molecular, Medewar Centre, Oxford Science Park, Oxford

[4] Daylight Chemical Information Systems Inc, 27401 Los Altos, Mission Viejo, California 92691

Chapter 9

Design of Small Libraries for Lead Exploration

Per M Andersson,[1] Anna Linusson,[1] Svante Wold,[1] Michael Sjöström,[1] Torbjörn Lundstedt[2] and Bo Nordén[3]

1. *Research Group for Chemometrics, Department of Organic Chemistry, Institute of Chemistry, Umeå University, SE-904 87 Umeå, Sweden*
2. *Structure Property Optimization Center (SPOC), Pharmacia & Upjohn AB, SE-751 82 Uppsala, Sweden*
3. *Medicinal Chemistry, Astra Hässle AB, SE-43183 Mölndal, Sweden*

Key words: Combinatorial chemistry, PCA, statistical experimental design, QSAR, PLS

Abstract: A combinatorial chemical library is a (usually large) set of compounds made to contain all possible structures of a certain type. The library is often made in order to find a lead compound for a specific drug action or for the optimisation of a lead. Because of the large number of synthesised compounds in the library, their biological activity is usually measured by rapid and simple tests, *i.e.* "high throughput screening" (HTS), giving crude answers, for instance "active" or "not". Combinatorial chemistry (CombC) comprises a chain of parts linked by the objective of finding lead compounds for further development. Sometimes the objective is to optimise an existing lead compound, but this is not much discussed in this chapter. An analysis of this CombC chain indicates that the biological testing is the weakest part of the chain. This is due to the difficulty in performing an in-depth biological testing of any set of compounds exceeding a couple of hundred members. Hence there is a strong motivation to decrease the size of libraries to a size that allows in-depth biological testing.

We discuss how the size of a library can be drastically reduced without loss of information or decreases in the chances of finding a lead compound. The approach is based on the use of statistical molecular design (SMD) for the selection of library compounds to synthesise and test, followed by the use of quantitative structure activity relationships (QSARs) for the evaluation of the resulting test data. The use of SMD and QSAR is, in turn, critically dependent on an appropriate translation of the molecular structure to numerical descriptors, the recognition of inhomogeneities (clusters) in both the structural and biological data spaces, and the ability to analyse and interpret relationships between multidimensional data sets.

We present a strategy for constructing a library with optimal information while still taking synthetic feasibility into account. The objective is to provide optimal chemical diversity with a moderate number of compounds, plus adequate depth and width of the biological testing. The strategy is based on a multivariate characterisation of the synthesis starting materials (building blocks), Principal Component Analysis (PCA), multivariate design, and Multivariate Quantitative Structure-Activity Relationships (M-QSAR). The strategy applies both to solid phase synthesis as well as libraries in solution.

1. INTRODUCTION

Drug discovery is a complex process spanning a long period of time and involving many different chemical and medicinal disciplines. The whole process is dependent on the weakest or limiting step in the process (Figure 1). Previously the synthesis of new compounds and the biological testing was the limiting step. Combinatorial chemistry (CombC) changed the picture showing that it was possible to make a large number of compounds followed by rapid biological testing, *i.e.* high throughput screening (HTS) [1].

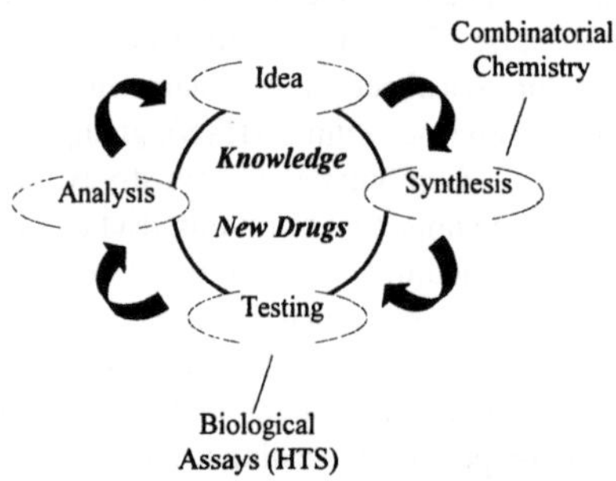

Figure 1. The drug discovery process.

In the initial phases of a drug design project, fairly large numbers of compounds, usually in the order of a couple of hundreds to a couple of thousands, are synthesised and tested [2, 3, 4]. With CombC, these numbers may now be increased by several orders of magnitude. It is not unusual to make and test millions of compounds in a single project [5]. CombC started about 10 years ago with the work of Geysen [6] and Houghten [7], who showed that, in the synthesis of peptides, one may make all or almost all of the possible compounds, and then find the ones that are most active by using sensitive and selective tests. This generated an avalanche of efforts to use the same principles to generate and test candidate structures (or mimics) for the purposes of drug design. More recently this has been extended to the synthesis of ordinary organic compounds of the type A-B-C. Diverse sets, or libraries, of building blocks (BBs) A_i, B_j, and C_k (i=1,2,..,I; j=1,2,..,J; k=1, 2,.., K) can generate a large set of combinations A_i-B_j-C_k with the help of automated parallel organic syntheses. The number of building block libraries (BBLs) is not restricted to three; the same approach can be used to make compounds of the type A-B-C-D, A-B-C-D-E, *etc.* Another approach for generating organic libraries is to combine a rigid core molecule supporting multiple reactivity sites, with a mixture of building blocks, which generates random libraries [8]. These approaches of rapidly and semi-automatically creating and testing large "libraries" of compounds has been labelled "combinatorial chemistry", here abbreviated CombC.

Put somewhat crudely, the philosophical basis of CombC is the belief that making and testing a larger number of compounds leads to a higher chance of finding an active compound than if a smaller number of compounds were investigated. In the beginning, CombC was believed to be a route to all possible compounds of a certain structural type. This would, with appropriate biological testing, of course, maximise the probability of finding the most active compound. However, we are now starting to realise that making all compounds is not possible, neither in principle nor in practice; chemistry is open ended. Suppose one set out to make a peptide library with the sequence length of 23 amino acids, using only the natural amino acids as building blocks. This would result in 23^{20} different possible combinations to make. Synthesising 1 mg of each would result in a library with a total weight of 1.7×10^{18} tons, which is about 30 times the weight of the planet earth [9].

The crude biological testing of a large library by HTS also presents an even more serious problem, namely that of the validity of the results for further drug development. Making just one or a few *in vitro* measurements, *e.g.* the binding to a few proteins, corresponds to the assumption that these *in vitro* measurements are strongly correlated to *in vivo* and clinical effects. This assumption is rarely tested, and hence the biological testing by HTS is a very weak link in the CombC strategy. A large number of compounds also present administrative problems.

Hence, continuing to use CombC as a route to very many compounds now corresponds to the assumption that each additional compound provides some additional information about the structural requirements for a successful drug candidate. A million compounds are presumed to give substantially more information than, say, a thousand compounds.

It is easy to understand, however, that there is no direct and strong relation between the number of investigated compounds, *N*, and the chance of finding a compound with substantial biological activity. Such a relation holds only when (a) all compounds are well spread out over the pertinent abstract structural space, S-space, and (b) the compounds are adequately tested. Spreading compounds over the available S-space corresponds to the selection according to a good statistical molecular design (SMD) for selecting combinatorial libraries and BBLs [10-13]. Thus, it matters a lot which compounds are made. It may be more informative to make and test 25 well-selected compounds than 25 000 poorly selected ones (Figure 2).

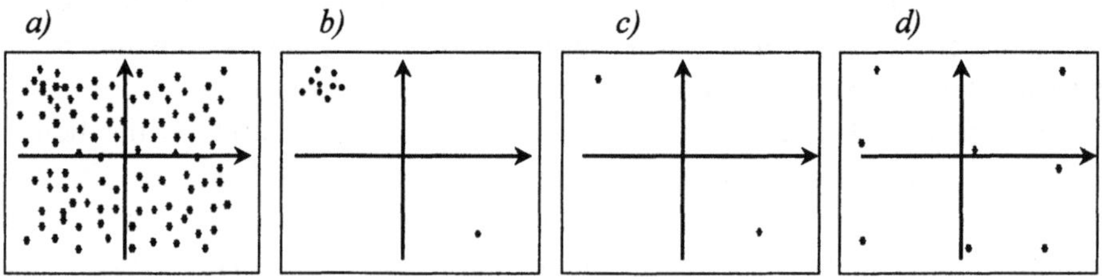

Figure 2. A compound can in a very simple case be represented as a point in an abstract structural space (S-space) where the two co-ordinate axes may, for example, represent pK_a and molecular weight. A set of compounds can then be represented as a set of points in the corresponding S-space. If a selection is made from a larger number of compounds *a)* the information content of the selected set is closely related to how well the set is spread out over the available region of the space. *b)* The information content in this set of nine compounds corresponds approximately to the same information as two compounds, as shown in *c)*. *d)* The nine compounds shown here constitute an informative set; each compound is far from the others, and together they span the investigated S-space well. In reality, the S-space is usually high dimensional with, typically, 20 to 30 axes.

The purpose of SMD in this context is to find an appropriate set of compounds, that can be synthesised by the fast and automated methods of CombC, at the same time as this set is optimally informative about the investigated biological activity and its structural requirements. The purpose of the modelling is to utilise this information efficiently for understanding and interpreting the relationship between structure and activity for interesting classes of compounds and using the resulting models for predicting the activity of new, yet untested, compounds, thus resulting in compounds with improved activity profiles.

2. COMBINATORIAL CHEMISTRY FOR OPTIMISING A LEAD

One way to use CombC is for optimising lead compounds. For this purpose we can use our experience of how to investigate congeneric series and how to develop Quantitative Structure Activity Relationships (QSARs) in order to increase our understanding and optimise structures. Here CombC is a practical approach to generate rapidly substantial sets of compounds that cover well the S-spaces of limited structural variation. The investigation and statistical modelling of such a set is needed for the purpose of optimisation. Of course SMD comes in here as an integrated part for the selection of these sets, as well as statistical modelling for connecting the results of all the tests of the compounds. How to do this is fairly well known and understood [3, 4]. The theme of this chapter is, however, the use of CombC for *finding* leads, as discussed below.

3. DEFINING A STRATEGY FOR LEAD EXPLORATION

3.1 Chemometrical Aspects

The CombC approach implicitly or explicitly involves several types of modelling. Biological models are used for the biological testing; chemical, statistical and chemometrical models are used for selecting compounds, and for relating the variation of the biological data to the variation of chemical structure. These models are employed even if we do not explicitly mention them.

The use of modelling is based on a number of assumptions. The most important relate to the "similarity" of the modelled set of phenomena or events or cases (here chemical compounds). Any mathematical/statistical, chemical/biological modelling approach must either be based on fundamental "*ab initio*" theory without many adjustable parameters, or be derived as a "Taylor or other serial expansions" of such fundamental models. In the latter case the model is valid only "locally", *i.e.* for moderate changes in its variables. In the present context, this corresponds to having only a moderate variation in chemical structure between the compounds included in one model, as well as only a moderate variation of the biological mechanism of "action" of the chemical compounds. If there is a wide variation, this necessitates the use of several models, one for each structural and biological-mechanistic "type" of compounds.

Since drug design and CombC involve the modelling of the relation between chemical structure and biological activity, it is essential that the investigated compounds span the pertinent space representing the structural variation in a reasonable way. This leads to the use of SMD for the selection of these compounds, as further discussed below.

CombC can be used as an approach to optimise a known drug candidate, *i.e.* a lead compound, as well as for finding new lead compounds. This involves more than just making many compounds, namely;

1. How to describe the chemical structures quantitatively.
2. How to select which compounds to make.
3. How to make the biological testing.
4. How to evaluate the test results.

These problems will be discussed below.

3.2 Defining the Search Space

To be able to compare quantitatively chemical compounds with respect to similarity and diversity, and to model the relation between their structural variation and biological activity, we need to translate the chemical structure (including conformation, tertiary structure, *etc.*) to a descriptive array of numbers. This translation must be uniform, *i.e.* each compound must be described by the same set of variables and the variables must mean the same thing for all compounds. This is an area where great progress has been made recently with computational chemistry and physical organic chemistry providing the essential tools and concepts [3, 4, 10, 12]. Usually we can arrange this array of numbers as a vector, where each vector element of one compound corresponds to the same element of another compound. We call these vector elements structure descriptors, or structure descriptor variables.

This quantitative description has a number of requirements to be suitable for CombC. It has to be fast to derive or calculate. It has to be relevant, *i.e.* capture the essential structural properties that influence the biological activities of interest. The structure descriptor variables have to be chemically interpretable. If possible, the description should be reversible, *i.e.* be possible to also translate backwards from descriptor values to structure. The description should be consistent with similarity ↔ dissimilarity, so those compounds that are closely similar have very similar values for the descriptors, and dissimilar compounds have widely different values for at least most of the descriptors.

There are two choices for this structure quantification, either the structural description of the building blocks (BBs), *i.e.* the starting material, or the description of the whole final library of molecules, *i.e.* the products. The former has several advantages;

- Describing the separate BBs, *e.g.* A_i, B_j and C_k, indirectly gives a description of their combinations, $A_iB_jC_k$
- Smaller structures result in simpler molecular calculations for characterisation
- Literature data are often available
- Fewer compounds are needed to be characterised with BBs compared with all combinations
- Characterised BBs can be used in more than one project/library
- Describing the separate BBs also leads to the possibility of using the descriptors for making a selection of BBs according to an SMD.

Because of these advantages only this alternative is considered.

Physical and computational chemistry provides us with good approaches for quantification of structure. These apply differently for the two cases of a BBL being a fixed backbone (scaffold) plus "substituents", or when the BBL has a more complicated composition.

(a) Classical substituents. Over the last six decades physical organic chemists have developed a number of structural descriptors – substituent scales – for the parts of organic compounds that can be seen as varying around a fixed backbone, the organic substituents. These scales describe electronic (Hammett, Taft), steric (Taft, Verloop), lipophilic (Hansch, Rekker), and hydrogen bonding effects [3, 14-16].

For an early discussion of the use of this type of descriptor combined with statistical experimental design, see Hellberg *et al.* [17] More recently, for the design of peptide libraries and QSAR modelling, principal properties have been described for a total of 87 natural and unnatural amino acids [18-20].

(b) Other fragments. Often BBs cannot be described in terms of a common backbone with varying substituents. One then needs a way to calculate structure descriptors that still capture steric, lipophilic, and polar effects. Here 3D approaches, *e.g.* the GRID approach [21], seem to work well, accounting both for intrinsic properties such as volume, shape, atom type, *etc.*, as well as interactions with the surroundings of the molecule (van der Waals, hydrophobic, Coulombic, and hydrogen bonding). A recent example of this approach is given by the GRID characterisation of heteroaromatics [22].

For the description of large numbers of molecules needed for selecting a BBL, however, 3D and other computational chemistry methods tend to be slow and cumbersome involving conformational problems. They need to be automated and connected to other software for the automatic generation of structural co-ordinates. The development of such systems is underway in several laboratories, but they are not yet routinely available.

3.3 Pre-Processing of Structural Data using Principal Component Analysis (PCA)

In the common case when the K structural descriptors are not independent, but are strongly correlated, an initial principal component analysis (PCA) [23] should be made of the descriptor data to find the "real" dimensionality of the data. PCA summarises the largest variance in the data with principal components, denoted t_i. The new generated variables (t_i-scores) are here called principal properties. These variables are suited as design variables because they are independent, few and contain most of the systematic information from the data from which they were generated. Once the PCA has been performed, the design is made on the resulting principal components (PC) scores – so called multivariate design [24-27]. It is important to remember that one should inspect the score plots (t_i) and remove outliers and other compounds that are extreme. A rough guide for creating a robust PCA for use in design is to delete, say 5-10%, of the structures in each score (t_i). The reduced set of compounds is then used for further design calculations.

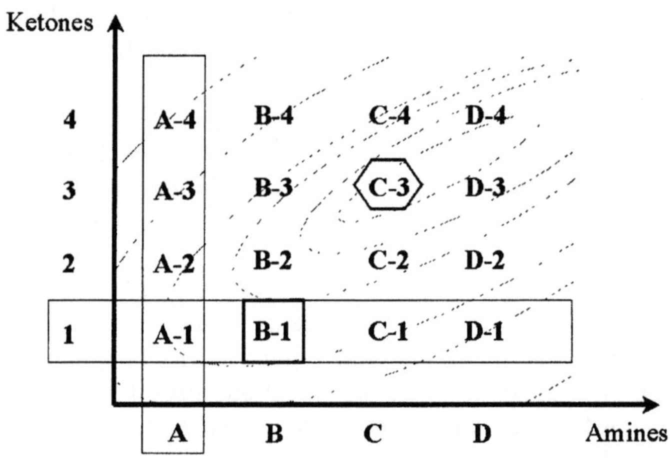

Figure 3. Map of a theoretical biological activity. The biological activity has been contoured against the structures arising from a particular combination of amine and ketone fragments. If at first the amine fragment is held constant (amine **A**) the ketone associated with maximal activity is **1**. If the ketone is held constant as **1**, the most potent amine is **B**, resulting in the false optimum **B-1** (in square). The true optimum **C-3** (in hexagon) is never found.

The aim is to find a drug lead or a drug from a drug lead, by optimising the structure toward some biological receptor. Often the mechanisms of binding to a biological receptor are unknown and the optimal structure has to be found with empirical methods. Optimal biological activity usually depends on several factors. If a structure type under investigation has two possible substituents that can be optimised, they have to be varied together. If they are changed one at a time and there are interaction effects between the substituents, a false optimum will be the end point (Figure 3).

Therefore it is necessary to use methods which allow all variables (substituents) to be investigated simultaneously, *i.e.* multivariate methods [26]. This can be achieved by using design (*e.g.*, factorial design [28] or D-optimal design [29]) in principal properties and evaluating the result by multivariate analysis as shown below.

3.4 Design in Principal Properties - Selection of Building Blocks in Clusters

From a statistical experimental design point of view, there are at least two ways to select compounds, (a) the design can be performed in the total space, *i.e.* when all the combinations have been computer generated, or (b) the design can be applied to the sets of building blocks. There are a number of advantages in using the latter approach. As mentioned before, it is very time consuming to achieve a good characterisation of the compounds in the first approach. Also, with respect to the synthetic aspects of CombC, it is actually a reduction of different building blocks that is desired, which of course is more efficiently accomplished by a design in the sets of building blocks.

When using CombC the resulting compounds will often be clustered. This is due to the fact that the building blocks are combined in a systematic way. A poor design in the building blocks can therefore cause a large area in the total structural space not to be covered. Consequently, the sets of building blocks have to be investigated for clusters and when the SMD is applied, D-optimal designs for models with linear and quadratic terms should be used [30].

An example using ketones, which are potential starting materials, *i.e.* a set of BBs, is presented below. We may imagine that we wish to make a library from A_i = ketones (R_1COR_2) reacting with B_j = Grignard reagents (R_3MgCl), then further reacting with C_k = acid chlorides (R_4COCl), to give the final library $A_iB_jC_k = R_1(R_2)(R_3)C\text{-}OCO\text{-}R_4$. A number of ketones were characterised multivariately and modelled by Carlson, Prochazka and Lundstedt [31] The structural variation is approximately captured by two principal properties, denoted by t_1 and t_2, as indicated in figures 4 and 5. The Grignard reagents and acid chlorides, both being of the generalised structure R-X, are assumed to be well characterised also by two principal properties, which will not be further considered in any detail.

Figure 4 shows the distribution of the ketones in the two dimensional score space (t_1, t_2), resulting from the principal component analysis (PCA) of the table of 78 ketones described by the 11 structure descriptor variables derived from IR, NMR spectra and other properties such as density, molecular weight and so on [31]. The figure also shows 9 compounds selected by a D-optimal design to well span this score space. Figure 5 shows the same score space but with another selection of 12 compounds, claimed to be superior.

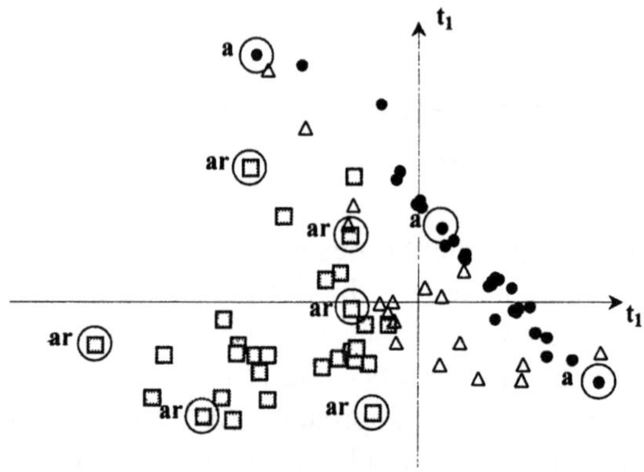

Figure 4. An S-space defined by the two first principal component scores (principal properties t_1 and t_2) of the 78 ketones. The rings indicate the selection of 9 compounds according to a D-optimal design and a quadratic model in t_1 and t_2. The design has selected three aliphatic ketones (prefix a, filled black circles), and six aromatic ketones (prefix ar, filled grey squares). No alicyclic ketones (open triangles) were selected by the design.

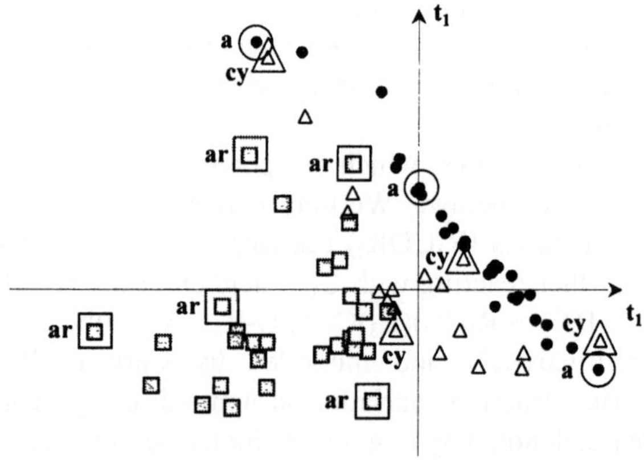

Figure 5. The same S-space as in Figure 4, with aliphatic (prefix a, filled black circles), aromatic (prefix ar, filled grey squares), and alicyclic ketone (prefix cy, open triangles). The

selection of compounds was made according to three separate D-optimal designs for the three classes, with three, four, and five compounds for a, cy and ar, respectively.

A set of building blocks (a BBL) may be clustered, *i.e.* the S-space contains dense regions of structures, clusters, with empty space in between. If things are done properly each cluster will contain similar compounds. In this case, a separate SMD made for each cluster will provide an adequate representation of the whole S-space [32]. If there are compounds between the clusters, a separate designed selection of these should be made to complement the selection of BBs.

Looking at the ketones, there are examples of clustering. The clusters can also be interpreted in a chemical sense, *i.e.* the clusters consist of aliphatic, aromatic and alicyclic structures. A design in the PC scores (N=9, D-optimal with center point, for a quadratic model) made without the recognition of the clusters is shown in figure 4. Although the design covers the space of t_1 and t_2, it completely misses the alicyclic compounds. Also, the aromatic group is over-represented due to its greater area in the PC plot. Instead the design should be made in each separate cluster (Figure 5). These designs were made with five points for the aromatic group, four for the alicyclic, and three for the aliphatic group. This selection is much better since it has an adequate coverage of the three types of ketones, with about the same total number of compounds. Adding, say, 6 randomly selected compounds gives a BBL of ketones containing N=18 compounds. A set of randomly selected compounds should always be added to the selection made by design in a PCA model. This is because all models are incorrect and we do not know every thing, therefore the designed selection should be complemented with a random selection from the BB candidate set.

Assuming (without solid motivation) that the Grignard reagents and acid chlorides have 4 and 5 clusters, similar reasoning would give BBLs of around 25 and 30 compounds, respectively, for these classes.

We call this approach cluster design, where we first divide each set of BBs into subsets (clusters) of similar compounds, then making BBLs by separate SMDs for each cluster. The approach has positive consequences: Simple designs result, because structural variation in each subset is limited. Relevant descriptors are not too hard to find because subsets can often be seen as backbone + substituents, and we have lots of good substituent descriptors for *e.g.* size, lipophilicity, polarity and polarizability. Finally, QSARs of each cluster are possible to construct for the same reasons, making results interpretable and predictions possible.

When applying cluster analysis for finding groups, we must remember that clusters are rarely spherical in S-space, but elongated as seen in figures 4 and 5. Also the different clusters have different variance-covariance structure, *i.e.* different orientation in S-space. Cluster analytical methods that are consistent with this non-roundness and differences in orientation include those of fuzzy clustering [33] and SIMCA [34, 35]. Also hierarchical structures are common in chemistry, *i.e.* grand clusters divided in sub clusters, divided in sub clusters, *etc.*

Often, performing PCA on a BBL indicates that the S-space is clustered – exemplified by the ketones. More examples can be found in the literature, *e.g.* the heteroaromatics [22] (10 clusters) and alcohols [36]. A second separate PCA for each cluster may be needed to derive good principal properties, design variables, for the clusters. This was not done here in the ketone example, however.

3.4.1 Design of the Building Block Combinations

Once a selection of BBs has been performed in the principal properties, the structural spaces of the BBs are well covered. By combining all of the selected BBs, the number of possible products in the designed library is given. The selected combinations are a good design to cover the structural space in the final library. However, if it still results in many combinations, too many to be synthetically, practically or economically feasible, another selection/design must be performed. Our approach is then to make a new design based on the selected combinations. By using fractional factorial designs [26] the number of compounds can be reduced (Figure 6). In this selection, synthetic feasibility and BBs cost and availability can be taken into consideration, *i.e.* if one combination is suspected of being difficult to synthesise, another combination can to some extent be chosen.

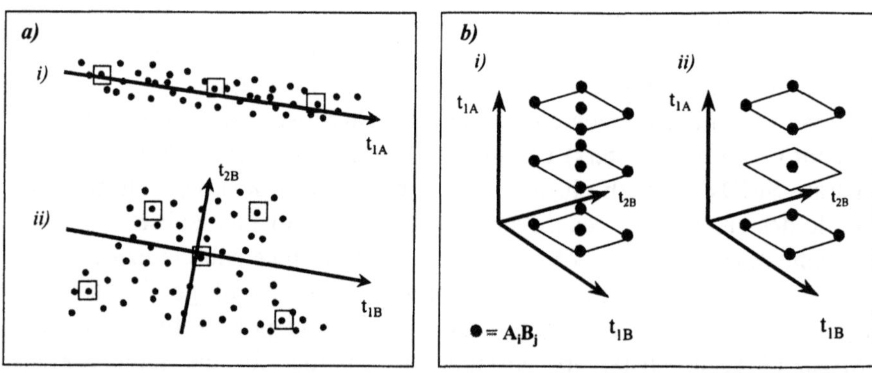

Figure 6. a) Design in principal properties. *i)* BBs of type A is described with one principal component. Design in A - a selection of three BBs that cover the structural space of A is made. *ii)* BBs of type B is described with two principal components. Design in B - a selection of five BBs that cover the structural space of B is made. b) Combination (A_iB_j) of selected

building blocks (A_i and B_j). *i)* A full factorial design, *i.e.* all combinations of the selected building blocks. A good design, but it requires $N = 15$ experiments. *ii)* A fractional factorial design with $N = 9$ experiments. A reduced design that still contains a maximum amount of information about the investigated structural space A_iB_j.

For more complicated libraries, when there are more than two building blocks which result in separate designs in each, designs other than fractional factorial designs could be used, *e.g.* D-optimal designs.

3.4.2 A Design Example

A synthesis was planned between commercially available primary aliphatic amines and aromatic aldehydes as building blocks. 35 amines were characterised with 11 physico-chemical measured and calculated variables [37], and 44 aromatic aldehydes were characterised with 54 generated variables [38]. A PCA of the amines resulted in a one component model which explained 72% of the variance in X. Three compounds were selected that covered the S-space of the primary amines (Figure 7a). Investigating the order of compounds from low to high t_1 –value shows that the major reason for the separation of the amines is size. The compound with the lowest t_1-value is methylamine (C_1), the highest is dodecylamine (C_{12}) and in between there are amines of various sizes and branching of the carbon chain. Another PCA was performed on the aldehyde data set. This resulted in a three component model which explained 88% of the variance in X. A selection of five compounds from these principal components covers the S-space of the aromatic aldehydes (Figure 7b). SIMCA software [39] was used to perform the PCA. An eigenvalue larger than two was used as a significance criterion for the extracted components.

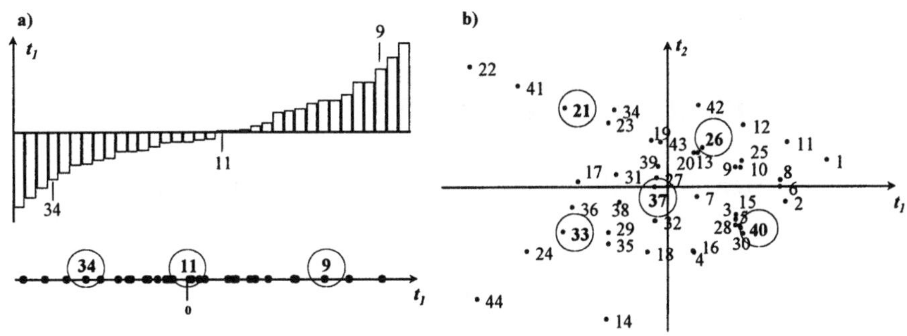

Figure 7. Principal Component Analysis (PCA) for the two building blocks. a) The first and only principal component (t_1) for the primary amines are illustrated with a bar diagram and a t_1 plot. The three selected amines are numbered and shown encircled. b) Plot of the first two

principal components (t_1 versus t_2) for the aromatic aldehydes. The encircled points with bold numbers are the five selected aldehydes.

All possible combinations between the original amine and aldehyde data sets would result in 35×44 = 1540 products. By using SMD a subset of 8 structures has been selected (Figure 8). Making all the combinations of the three amines with the five aldehydes would result in 3×5 = 15 products. These 15 compounds are a good design for covering the investigated S-space with only a few experiments. If there is a constraint (*e.g.* cost or time limit), a reduced design where only nine compounds are selected can be made, using this first design as a basis, as described above.

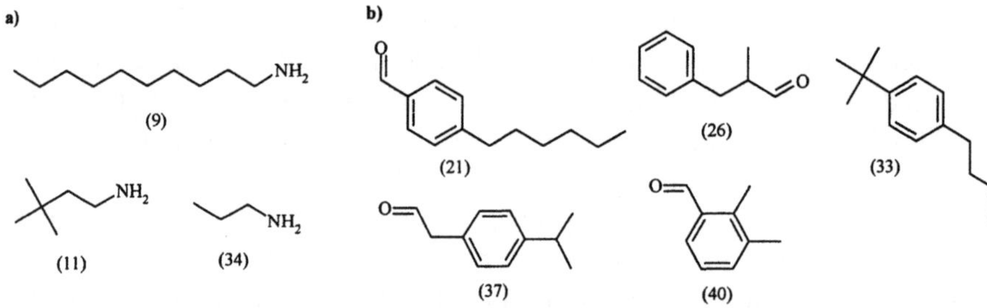

Figure 8. The structures selected with SMD in each structural subspace. a) the 3 chosen primary amines; b) the 5 chosen aromatic aldehydes. The numbering is the same as in figure 7.

The same approach can be used for larger data sets and will be investigated further [30]. Important is that clusters must be identified and separate designs made within each cluster. In this example, primary amines and aromatic aldehydes were used and no apparent groupings were observed within these classes.

4. CHEMICAL SYNTHESIS

The first step in the chemical synthesis is to investigate the scope and limitations of the synthetic reactions [32, 40], since different clusters of compounds, or even compounds within a cluster, may require different experimental settings. Even with automated solid state syntheses using polymer beads as carriers of compounds, the synthetic steps need to be optimised so that reasonable yields are obtained for all compounds in the library. Otherwise unbalanced test data will result, with subsequent loss of information. Such optimisation is easy to accomplish with few compounds. Recently, robots that optimise organic synthèses on the basis of statistical modelling and experimental design have been developed [41]. For a moderate number of compounds, say up to 1000, the robots can be programmed to perform the entire synthesis, including the optimisation of each individual step, in a reasonable time. These possibilities are another strong argument for limiting the number of compounds and for just making as many as needed from the information point of view.

5. BIOLOGICAL TESTING

5.1 Importance of Good Biological Testing

The biological testing is the most important step of the CombC as well as of any other strategy for drug design. Without good and relevant measures of the biological activity, no rating of compounds can be made, and nothing sensible is achieved regardless of how many compounds are made. Hence we consider the biological testing to be an integral part of CombC, not a separate step glued on afterwards. Since the information content of the biological test data is so important, we strongly recommend moderate throughput multiple model screening (MTMMS) instead of high throughput screening (HTS) which is usually favoured in CombC applications.

In the pharmaceutical industry, the objective is to develop drugs with desired clinical effect(s). To achieve this it is necessary to have access to tests, which are good model for the clinical situation. Hence these test systems are often called pharmacological models. These models have very different degrees of sophistication, from simple and fast *in vitro* models to very complex and time-consuming *in vivo* models. They have, as well, a large variation in relevance and adequacy, ranging from being good to very poor approximations of the clinical situation.

A fast biological screen will necessarily be only a crude approximation of the clinical situation. To get useful biological data it is still necessary to focus on the (often hypothetical) biochemical mechanism explaining the reasons for a disease. The mechanism has to be expressed in biological activities, which can be measured in HTS, *e.g.* receptor affinities and enzyme inhibition.

In CombC, it is common to apply HTS in order to find compounds with a specific activity, *e.g.*, compounds which are selective enzyme inhibitors. The HTS is then run only to trace active compounds (not selective) by just giving an active/inactive classification against the target enzyme. However, if one is to find and pursue any opportunities to "optimise" the selectivity, it is necessary to measure on a wider spectrum of fast screening models.

Hence, in order to obtain the best possible information from CombC, several HTS models must be used in parallel, *i.e.* MTMMS.

5.1.1 Risk of False Biological Test Result

Fast shallow testing of many compounds involves two types of risks, that of false positives (over-optimistic evaluation), and that of false negatives (missed opportunities).

5.1.1.1 Risk for False Negatives

With a large library of compounds, the only biological testing that can be afforded is quick and shallow "high throughput screening", HTS. The "hits" in this screening process are then taken to further testing with more informative, but also more time consuming, tests. This strategy involves great risks. It works only if the initial HTS results are strongly correlated to the "real" biological activity. Of course this can rarely be investigated beforehand, one just has to assume that a simple test of receptor or enzyme binding is informative about this "real" activity.

The risk for false negatives, *i.e.*, that an actually "active" compound is classified as inactive is a property of the actual biological test, and the only way to decrease it is to use more extensive, more deep biological testing. This takes time and resources, and hence the number of compounds in the library must be weighed against the demands on resources for the biological testing.

5.1.1.2 Risk for False Positives

The larger the number of tested compounds, the greater is the risk of false positives for a given biological test. This occurs because each compound is evaluated separately, with a given probability that the test result is falsely positive. Hence, CombC with many compounds evaluated by HTS may give a large number of false positive compounds.

The risk for false positives when evaluating a large set (with N members) one compound at a time increases as $\{1 - 0.95^N\}$, if we consider compounds having an activity larger than 2 SDs (standard deviations of the testing variability or $\{1 - 0.99^N\}$ for activity more than 3 SDs away. With the usual testing noise of 0.3 log units (a factor of two in potency) the Bonferroni correction should be applied to avoid too many false positives, *i.e.*, testing on the probability level of $0.05/N$ or $0.01/N$ instead of 0.05 or 0.01 respectively. With large values of N, this leads to very small probability levels, and absurd requirements on activity levels to be accepted as a "real hit". This is another strong argument for (1) having moderate values of N, and (2) evaluating all compounds together with respect to some kind of model, distribution and/or structure related, to get the problems of false positives under control.

Another more problematic type of a false positive is a compound with real activity in the target system, but which for some reason is useless for further development. This may be, for example, a compound that binds well to a receptor, but which is metabolised by other enzymes in the cell or on the way to the "target" and this is not detected by shallow biological testing. The best way to circumvent this problem is to test fewer compounds and to have a more extensive biological testing including, *e.g.* tests related to metabolism and reactivity. This will lead to a much smaller risk for a false positive result. Again, we see the need to make a compromise between the number of compounds and the depth of the biological testing.

A similar argument applies to false negatives. Since any single test value has a non-negligible measurement error (noise), there is a certain risk of missing real positive results. A good way to decrease the total risk is to analyse all data together by means of a model, *e.g.* a multivariate QSAR model, were all activity values support each other and the consequences of noise and measurement errors are minimised.

5.1.1.3 Depth of Biological Testing.

To avoid or reduce these problems with false negatives and positives, the following is suggested;

1. Run several HTS models in parallel.
2. If possible, investigate all of the biological models with a set of compounds with known clinical activity, designed in accordance with a multivariate design accounting for the clinical action as well as the chemistry.

By doing this it is possible to make a multivariate analysis of the results and obtain information on the discrimination power of the different HTS models.

Hence there are strong arguments for an as extensive biological testing as possible. This is, in turn, a strong argument for the use of SMD which may greatly reduce the number of compounds needed to give a certain amount of information. Hence, with appropriate SMD the compounds may be tested in more depth, thereby improving the total, chemical *and* biological, information. We shall call this testing strategy MTMMS for moderate throughput multiple model screening.

5.2 Analysing the Biological Result – Multivariate QSAR Modelling

When compounds are selected according to SMD, this necessitates the adequate description of their structures by means of quantitative variables, "structure descriptors". This description can then be used after the compound selection, synthesis, and biological testing to formulate quantitative models between structural variation and activity variation, so called Quantitative Structure Activity Relationships (QSARs). For extensive reviews, see references 3 and 4. With multiple structure descriptors and multiple biological activity variables (responses), these models are necessarily multivariate (M-QSAR) in their nature, making the Partial Least Squares Projections to Latent Structures (PLS) approach suitable for the data analysis. PLS is a statistical method, which relates a multivariate descriptor data set (X) to a multivariate response data set Y. PLS is well described elsewhere and will not be described any further here [42, 43].

The use of QSARs for the final evaluation of the test data further decreases the risk for false positives and false negatives. This is because all data are analysed together, and thereby the random variation of individual test results are averaged out. Moreover, outliers and erroneous test results are identified by model diagnostics, which further stabilises the results.

5.2.1 M-QSAR Example

An example is presented to illustrate M-QSAR. The example is taken from the literature, where carboxylic acids were tested against the ability to cause skin corrosion in rabbits [44]. A set of 45 carboxylic acids were characterised with nine different calculated and experimental properties. The biological response was expressed as lowest observed effect concentration (LOEC). A PCA was performed on the X matrix (45×9), which resulted in a three component model according to cross-validation with seven cross-validation groups. The three new orthogonal variables for the candidate set, *i.e.* the principal properties, were used to select a training set of nine compounds and a test set of six compounds according to an approximate fractional factorial design [45]. The training set is composed of the structures (here acids) which are the basis for the biological model and the test set is used to validate the generated model. The design of selected compounds is illustrated in figure 9.

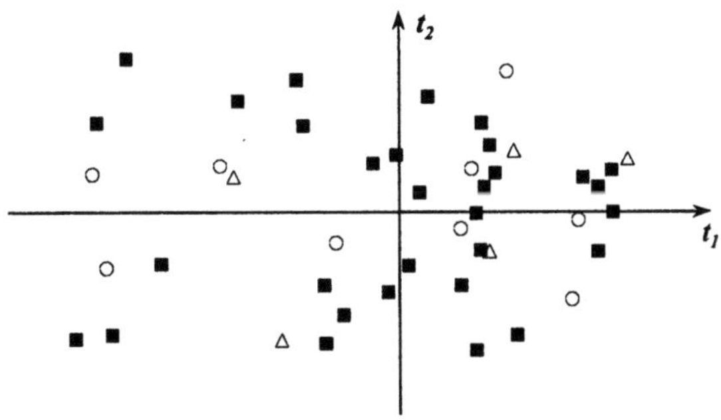

Figure 9. Score plot, t_1 versus t_2. The factorial design in the principal properties of the 45 carboxylic acids. Open circles are the training set and are the basis for the M-QSAR model. Open triangles are the test set used to validate the model and filled squares are untested acids.

It is important to notice that these 45 acids form a homogeneous set, *i.e.* there are no strong groupings or outliers, in the score plot. A total of 15 acids were selected and tested. LOEC was determined for 14 of the 15 acids. For one of the acids LOEC could not be determined, because it was not corrosive in the investigated concentration interval. This resulted in a training set of nine acids and a reduced validation set of five acids. From the training set, a model was calculated using PLS. In order to obtain a good model, the X matrix was expanded with the quadratic terms of the nine descriptors. This was done to model non-linearity between the X and the Y. The generated two component PLS model explained 87.5% of the variance in Y, *i.e.* R^2Y_{cum} = 87.5%, and had a predicted capability Q^2_{cum} = 60.7% based on cross-validation with nine cross-validation groups. To test the model further, predictions were made for the test set. The result shows that the generated model has a predictive ability (Figure 10), and can be used to estimate the corrosive ability of the untested carboxylic acids, as well as additional carboxylic acids similar to the 45 used in the candidate set.

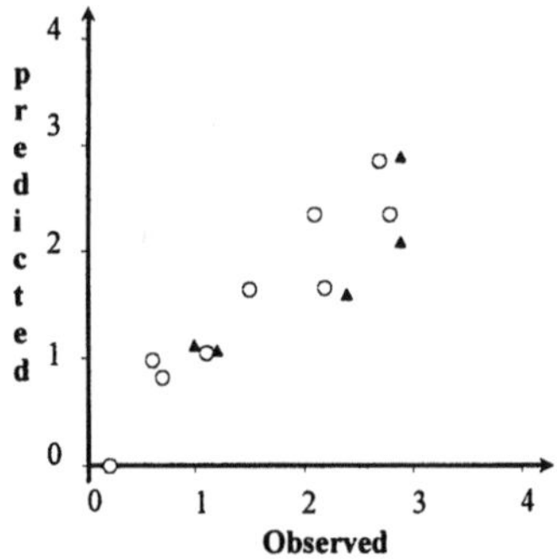

Figure 10. The M-QSAR model used to predict the biological response. Observed versus calculated for the training set (open circles) and predicted for the test set (black triangles).

6. DISCUSSION

The selection of compounds to be synthesised to form a library is made with the purpose of making the library as "informative" as possible with respect to a certain type of biological activity. The library compounds will be tested with either the hope of finding a "lead" for a certain pharmaceutical application, or for the optimisation of a certain class of compounds with respect to, usually, a profile of biological and physico-chemical measurements.

We have emphasised that the chances of having a real positive hit in at least one compound library are not directly related to the number of compounds in the set, but rather to how well the library covers the pertinent structural space.

The use of multiple designs and multiple models do not lead to any loss of information or understanding. It is just a consequence of our knowledge that for a model to have any validity, it must be based on, and derived from, a set of homogeneous data. Here this means a set of compounds and their biological activities where the interaction between structure and "biology" remains essentially the same. The same multivariate and interpretation tools that are used to find the "clusters" can be used to classify new compounds according to which cluster they are most similar. Hence predictions and optimisation are achieved as easily with multiple models as with a single one. Moreover, since the separate cluster models are much simpler than any "global" single model, the clustering also simplifies the interpretation and the understanding of the results. The idea of dividing structure space into clusters is a natural thing to do, considering that this has been done in organic chemistry for a very long time, *i.e.* organic, inorganic, alcohols, amines, primary, secondary, aromatic and so on.

An important part of the philosophy behind the proposed approach is the recognition of the importance of the biological test data. The drug discovery process is still limited by the quality of the biological test systems. Without relevant and fairly extensive biological test data nothing is achieved at all. This is a serious weakness with HTS data; the correlation between HTS results and real clinical activity is often weak or even absent. This makes CombC based on casual HTS equivalent to throwing dice but much more expensive. The biological testing must be an integral part of CombC, not a separate phase that can be glued on afterwards.

The same strategy can be used for lead optimisation as well. The main difference between lead generation and lead optimisation is that in the latter case, the class of structures is rather well specified. Only one or a few "clusters" are then investigated and thereby rather dense designs (RSM and/or space-filling designs) can be applied. For lead generation the class of structures is not very well known, and the search consists of investigating a larger number of clusters. The corresponding design is rather sparse, *i.e.* a screening design.

Comparing our proposed strategy for CombC with what actually is done in most laboratories today, there are differences in (a) the selection of compounds (we recommend the use of design, SMD), and (b) the evaluation of the results (we recommend a formal multivariate modelling similar or equivalent to M-QSAR). Martin, Blaney *et al.* at Chiron [10], and Young *et al.* at Glaxo[11], and others, use design. The design part is difficult to do well without the recognition of possible clusters in the data. In related areas such as selecting compounds for QSAR, for chemical synthesis, *etc.*, we have long stressed the necessity to model, and therefore design, clusters separately[42, 43].

ACKNOWLEDGEMENTS

Financial support from the Swedish Natural Science Research Council, Astra Hässle AB and Pharmacia & Upjohn AB to the Umeå Chemometrics Group are gratefully acknowledged.

REFERENCES

1. Broach, J.R., Thorner, J. High-throughput Screening for Drug Discovery, *Nature*, 1996, Suppl., **384**, 14-16.
2. Spilker, B. 'Multinational Drug Companies'. Issues in Drug Discovery and Development. Raven Press, New York, 1989.
3. van de Waterbeemd H. (Ed.), *QSAR: Chemometric Methods in Molecular Design. Methods and Principles in Medicinal Chemistry*, **2**, Verlag Chemie, Weinheim, 1995
4. Kubinyi, H. (Ed.), '*3D QSAR in Drug Design; Theory, Methods and Applications*'. ESCOM Science Publishers, Leiden, Holland, 1993.
5. Houghten, R.A., Pinilla, C., Bondelle, S.E., Appel, J.R., Dooley C.T. and Cuervo, J.H. Generation and Use of Synthetic Peptide Combinatorial Libraries for Basic Research and Drug Discovery. *Nature*, 1991, **354**, 84-86.
6. Geysen, H.M., Meloen, R.H. and Barteling, S.J. Use of Peptide Synthesis to Probe Viral Antigens for Epitopes to a Resolution of a Single Amino Acid. *Proc. Natl. Acad. Sci. U.S.A.*, 1984, **81**, 3998-4002.
7. Houghten R.A. General Method for the Rapid Solid-Phase Synthesis of Large Numbers of Peptides: Specificity of Antigen-Antibody Interaction at the Level of Individual Amino Acids. *Proc. Natl. Acad. Sci. U.S.A.*, 1985, **82**, 5131-5135.
8. Carell, T., Wintner, E.A., Bashir-Hashemi, A. and Rebeck, J.Jr. A Novel Procedure for the Synthesis of Libraries Containing Small Organic Molecules. *Angew. Chem. Int. Ed. Eng.*, 1994, **33**, 2059-2061.
9. Personal communication, T. Olsson, Astra Hässle AB, Sweden
10. Martin, E.J., Blaney, J.M., Siani M.A., Spellmyer, D.C., Wong, A.K. and Moos W.H. Measuring Diversity: Experimental Design of Combinatorial Libraries for Drug Design. *J. Med. Chem.*, 1995, **38**, 1431-1436.
11. Young, S.S. and Hawkins, D.M. Analysis of a 29 Full Factorial Chemical Library. *J. Med. Chem.*, 1995, **38**, 2784-2788.
12. van de Waterbeemd, H., Constantino, G., Clementi, S., Cruciani, G. and Valigi, R. Disjoint Principal Properties of Organic Substituents. In *QSAR: Chemometric Methods in Molecular Design, Methods and Principles in Medicinal Chemistry*, **2**, Ed. H. van de Waterbeemd, Verlag Chemie, Weinheim, Germany, 1995
13. Lundstedt, T., Andersson, P.M., Clementi, S., Cruciani, G., Kettaneh, N., Linusson, A., Nordén, B., Pastor, M., Sjöström, M. and Wold, S. Intelligent Combinatorial Libraries. In *Computer-Assisted Lead Finding and Optimization* Ed. H. van de Waterbeemd, Verlag Helvetica Chimica Acta, Basel, Switzerland, 1997, 191-208.
14. Hansch, C. and Leo, A.J. Substituent Constants for Correlation Analysis in Chemistry and Biology, Wiley, New York, 1979.
15. Hansch, C., Leo, A.J. and Hoekman, D. Exploring QSAR, Hydrophobic, Electronic and Steric Constants, ACS, Washington DC, 1995.
16. Skagerberg, B., Bonelli, D., Clementi, S., Cruciani, G. and Ebert, C. Principal Properties for Aromatic Substituents. A Multivariate Approach for Design in QSAR. *QSAR*, 1989, **8**, 32-38.
17. Hellberg, S., Sjöström, M., Skagerberg, B., Wikström, C. and Wold, S. On the design of multipositionally varied test series for quantitative structure-activity relationships. *Acta Pharm. Jugosl.*, 1987, **37**, 53-65.
18. Hellberg, S., Sjöström, M. and Wold, S. The Prediction of Bradykinin Potentiating Potency of Pentapeptides. An Example of a Peptide Quantitative Structure-Activity Relationship. *Acta Chem. Scand.*, 1986, **B40**, 135-140.

19. Jonsson, J., Eriksson, L., Hellberg, S., Sjöström, M. and Wold, S. Multivariate Parametrization of 55 Coded and Non-Coded Amino Acids, *QSAR*, 1989, **8**, 204-209.
20. Sandberg, M., Eriksson, L., Jonsson, J., Sjöström, M. and Wold, S. New Chemical Descriptors Relevant for the Design of Biologically Active Peptides. A Multivariate Characterisation of 87 Amino Acids. *J. Med. Chem.*, 1998, **41**, 2481-2491.
21. Goodford, P.J. A computational procedure for determining energetically favourable binding sites on biologically important macromolecules. *J.Med.Chem.*, 1985, **28**, 849-857.
22. Clementi, S., Cruciani, G., Fisi, P., Riganelli, D., Valigi, R. and Musumarra G. A New Set of Principal Properties for Heteroaromatics Obtained by GRID. *QSAR*, 1996, **15**, 108-120.
23. Jackson J.E. A Users Guide to Principal Components, Wiley, New York, 1991.
24. Wold, S., Sjöström, M., Carlson, R., Lundstedt, T., Hellberg, S., Skagerberg, B., Wikström, C. and Öhman, J. Multivariate Design. *Anal. Chim. Acta*,1986, **191**, 17-32.
25. Carlson, R. Design and Optimization in Organic Synthesis, Elsevier, Amsterdam, 1992.
26. Lundstedt, T. A QSAR Strategy for Screening of Drugs and Predicting Their Clinical Activity, *Drug News Perspect.*, 1991, **4(8)**, 468-474.
27. Sjöström, M. and Eriksson, L. Application of Statistical Experimental Design and PLS Modelling in QSAR. In *QSAR: Chemometric Methods in Molecular Design, Methods and Principles in Medicinal Chemistry*, **2**, Ed. H. van de Waterbeemd, Verlag Chemie, Weinheim, Germany, 1995
28. Box G.E.P. and Draper, N.R. Empirical Model-building and Response Surfaces, Wiley, Chichester, 1987.
29. Baroni, M., Clementi, S., Cruciani, G., Kettaneh-Wold, N. and Wold, S. D-Optimal Designs in QSAR. *QSAR*, 1993, **12**, 225-231.
30. A. Linusson, F. Lindgren, J. Gottfries, S. Wold, in manuscript
31. Carlson, R., Prochazka, M.P. and Lundstedt, T. Principal Properties for Synthetic Screening: Ketones and Aldehydes. *Acta Chem. Scand.*, 1988, **B42**, 145-156.
32. Lundstedt, T., Carlson, R. and Shabana, R., Optimum Conditions for the Willgerodt-Kindler Reaction. 3. Amine Variation. *Acta Chem. Scand.*, 1987, **B41**, 157-163.
33. Bezdek, J.C. Pattern Recognition with Fuzzy Objective Function Algorithms. Plenum Press, New York, 1981.
34. Wold, S. and Andersson, K. Major Components Influencing Retention Indices in Gas Chromatography. *J. Chromatogr.*, 1973, **80**, 43.
35. Dunn III, W.J. and Wold, S. SIMCA Pattern Recognition and Classification. In *QSAR: Chemometric Methods in Molecular Design, Methods and Principles in Medicinal Chemistry*, **2**, Ed. van de Waterbeemd, H., Verlag Chemie, Weinheim, Germany, 1995.
36. Linusson A., Wold, S. and Nordén, B. *Chemometrics and Intell. Lab. Syst.*, 1998, in press.
37. Carlson, R., Prochazka, M.P. and Lundstedt, T. Principal Properties for Synthetic Screening: Amines. *Acta Chem. Scand.*, 1988, **B42**, 157-165.
38. Tsar 3.11, Oxford Molecular Group, http://www.oxmol.co.uk/
39. Simca-P 3.01, Umetri AB, Umeå, http://www.umetri.se/
40. Carlson, R. and Lundstedt, T. Scope of Organic Synthetic Reactions. Multivariate Methods for Exploring the Reaction Space. An example by the Willgerodt-Kindler Reaction. *Acta Chem. Scand.*, 1987, **B41**, 164-173.
41. Scitec Laboratory Automation SA, Av. de Provence 18, CH-1007 Lausanne, Switzerland.
42. Wold, S., Johansson, E. and Cocchi, M. PLS - Partial Least-Squares Projections to Latent Structures. In *3D QSAR in Drug Design; Theory, Methods and Applications,* Ed. Kubinyi, H., ESCOM Science Publishers, Leiden, Holland, 1993, pp523-550.

43. Wold, S. PLS for Multivariate Linear Modeling. In *QSAR: Chemometric Methods in Molecular Design, Methods and Principles in Medicinal Chemistry*, **2**, Ed. van de Waterbeemd, H., Verlag Chemie, Weinheim, Germany, 1995, pp195-218.
44. Eriksson, L., Berglind, R. and Sjöström, M. A Multivariate Quantitative Structure-Activity Relationship for Corrosive Carboxylic Acids. *Chemometrics and Intell. Lab. Syst.*, 1994, **23**, 235-245.
45. Box, G.E.P., Hunter, W.G. and Hunter, J.S. Statistics for Experimenters, Wiley, New York, 1978.

Chapter 10

The Design of Small- and Medium-sized Focused Combinatorial Libraries

Design of focused combinatorial libraries

RICHARD A. LEWIS
Eli Lilly & Co. Ltd, Lilly Research Centre, Windlesham, Surrey GU20 6PH UK

Key words: SAR, library design, similarity, diversity.

Abstract: The use of focused combinatorial libraries is becoming an important weapon in lead exploration and optimisation. The challenge for the modelling community is to harness the experience and knowledge we have in generating structure-activity relationships (SARs) for use in library design. This paper describes the different methods of calculating similarity and diversity from an SAR, several strategies for library design, the interplay between the descriptors in the design process, and some practical examples of focused library design.

1. INTRODUCTION

One of the main challenges facing computer-aided drug design today is the design of combinatorial libraries. The vast numbers of compounds that could be generated by combinatorial synthesis are too large to be feasible using today's laboratory technologies, so some sensible design methods have to be used to reduce the virtual to the pragmatic. The main goal in the design of small and medium libraries is the rapid identification and optimisation of potent leads that have the potential to be clinically viable. In this chapter, the working practises, methods and the molecular descriptors that have been used in library design will be reviewed. Compromises have to be made between the rigour of the method and descriptors, the time taken to complete the design, the quality of SAR information being fed into the design, and the need to produce drug-like molecules. The bias of this chapter will be towards practical advice for designers and chemists, rather than towards theoretical analysis, which has been covered elsewhere [1].

In the early days of combinatorial chemistry, there was a tendency towards 'big is best', with libraries of many millions of compounds being synthesised and screened in mixtures. The rationale behind this was that the more compounds screened, the better the chances of finding a good lead for optimisation. In some cases, this method has brought success [2, 3], but it created other technical issues of mixture purity, mixture composition, and mixture deconvolution. Several ingenious solutions to these issues have been proposed [4], but current thinking within the pharmaceutical industry is tending to the view that the costs outweigh the benefits. Large libraries of single compounds are playing a valuable role in lead generation, but their synthesis has become the province of highly specialised labs. If a holistic view of high-throughput lead generation is taken, the bottlenecks have moved away from screening, synthesis and supply of compounds, and towards validation and optimisation. Small- and medium-sized libraries, made by Rapid Parallel Synthesis (RPS) techniques can provide the means whereby an SAR hypothesis or a lead can be validated quickly. This is the context in which library design takes place. It therefore follows that the rigour and effort that is put into a design should be matched to the resources available to make and test the library, the importance of the library to the SAR cycle, and the state of the optimisation cycle (that is, whether the lead needs radical changes or minor polishing). A library design which pushes the most promising compounds to the top of the synthesis and screening queue can offer substantial savings in discovery times [5].

1.1 Definitions

Combinatorial synthesis, or RPS, involves the combination of various reagents, according to a synthetic scheme, to generate products with one or more variable R-group positions (Figure 1). Assuming there are no limitations for synthetic reasons, all combinations of reagents at each of the positions may be generated. The size of the combinatorial library that could be generated is given by the product of the number of possible reagents at each of the substituent positions. For example, if a scheme involves three different reagents, and there are 100 possible reagents of each type, then the synthesised combinatorial library would contain 1 million compounds. A library can be built, or enumerated, in the computer, as a precursor to design and/or synthesis; the obvious term for this is a 'virtual library'. The number of compounds in a virtual library can be enormous; however, for the purposes of this paper, it is assumed that the library can be fully enumerated and scored. This limits the size to the order of 1 million compounds or less, depending on the speed at which they can be scored, relative to the time lines imposed by the synthesis team, and the number that can be loaded into memory for analysis. Scoring usually takes longer than analysis and design. If the design is required in 5 days, then the maximum size of the virtual library is approximately 5*24*3600/(the time in seconds taken to score 1 library member); this works out at 432000 if the scoring process takes 1 second per compound. If the virtual library has to be loaded entirely into memory for the analysis and design, then the maximum size of the virtual library is approximately (the amount of available RAM)/(the amount of information stored per library member). For more complex descriptors, the available memory is more often the limiting factor.

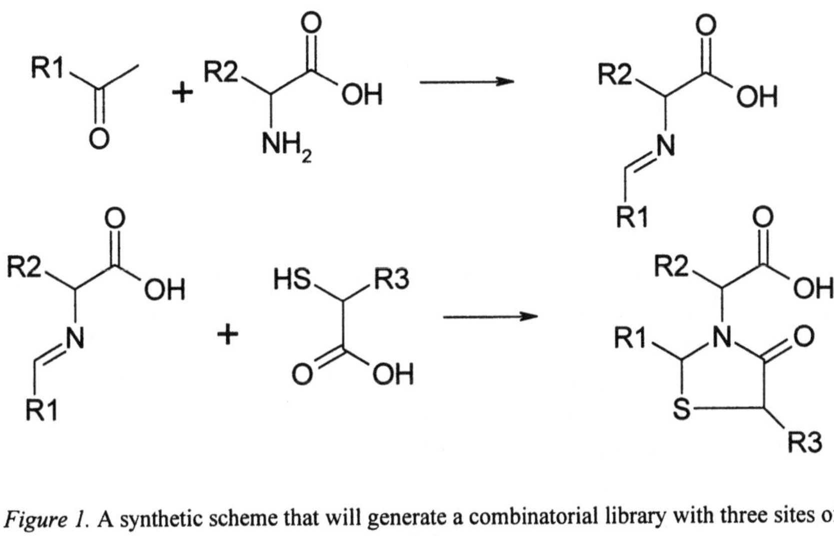

Figure 1. A synthetic scheme that will generate a combinatorial library with three sites of variation

The definition of what is a small or medium-sized library is fairly arbitrary, depending as it does on working practices within a laboratory. A team which places particular emphasis on purity may find that the rate-limiting step is the purification, rather than the actual synthetic steps. Similar constraints can be imposed by the complexities of the chemistries being attempted, and the final quantities required for screening. For the purposes of this paper, a small library will consist of only a few hundred members, while a medium library may have up to a few thousand members.

1.2 Combinatorial Efficiency

The minimum number of reagents needed to make N products in a k-component reaction is $k \bullet N^{1/k}$. The maximum number is k•N. Design methods that try to use the minimum number of reagents are called 'efficient', whereas those that tend towards larger numbers are termed 'cherry-picking'. The terms are not meant to be pejorative: the key factor in the design should be the exploration or refinement of an SAR, rather than the numbers of reagents used in the synthesis. Against that, it can be quite tedious and time-consuming to make a medium library that has been designed with no heed to efficiency. Thus, medium libraries will tend towards being truly combinatoric, that is, made up of all possible combinations of reagents, while small libraries need not be. This distinction is important, because it changes the design goal of maximising some measure of diversity and combinatorial efficiency to simply maximising diversity. In this latter situation, cherry-picking methods can be used. There is no universal recipe, and each case should be looked at on its own merits.

1.3 Diversity and Similarity

The terms 'similarity' and 'diversity' are very nebulous, as they seem to encompass several different concepts in the literature; Kubinyi has published an interesting polemic on this issue [6]. A narrow concept will be employed in this paper, which revolves around the context of exploring an SAR. Small and medium libraries are made for the purpose of following up leads and exploring their SAR as quickly as possible. The libraries therefore need to be designed around a hypothesis of what makes the leads active, and what factors that might contribute to the SAR which have not yet been explored. The library design, and hence the definition of diversity, must therefore vary from case to case. The logical conclusion of this line of argument would be exemplified by the design of peptides to bind to proteases: each design is based around a common core and a common chemistry, but each design will be different, driven by the environment of the different enzymes. One can make some general remarks about which descriptors will probably be important, and this will be covered later. Diversity is therefore the spread of observations in a defined descriptor space, and within defined limits of that space, the descriptors and limits being determined by the nature of the SAR, and the amount of knowledge available about the SAR.

1.4 Work Flows in RPS

A typical work flow for the conception, design and synthesis of a library is shown in figure 2. The starting point is SAR information and a synthetic plan for making the library. The first phase revolves around testing out the feasibility of the synthetic scheme and gathering information on the reagents available for use. These two processes can impose limits on the virtual library through providing constraints on which reagents will react, which reagents are available in the stockroom or by quick delivery. This leads into the reagent filtering phase and results in the final set of reagents for enumeration into the virtual library. The next phase is the design phase, which takes input from the SAR and other sources. Closely allied to the design phase is the inspection phase, in which the compounds chosen by the design are eyeballed by experienced medicinal and RPS chemists to make sure that the design meets their expectations. The next stages are synthesis, purification and registration, followed by screening and validation. If a library has been carefully designed according to an explicit hypothesis, then the results from biological screening should serve to test the hypothesis and guide the design of the next library. If the design has not been driven in this way, it will be that much harder to uncover the SAR information, thus defeating the object of making the library in the first place.

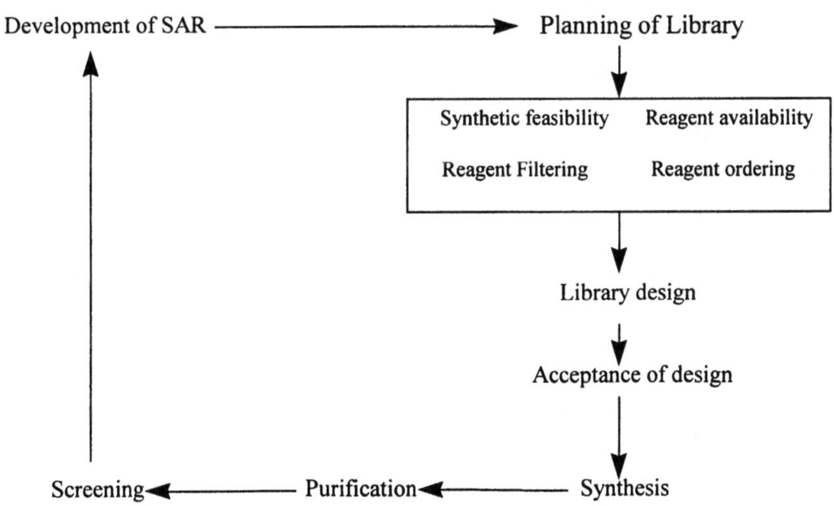

Figure 2. A workflow for the design, preparation and use of a combinatorial library

1.5 SAR information

SAR information can come from many sources, but the most common, certainly for initial (active hit) information will be high-throughput screening (HTS). In a good case, several close analogues (close in terms of substructural similarity) will have been found active, in a bad case the actives appear to be unrelated, and close analogues to the actives are inactive. A molecular modeller will try several approaches to deduce an SAR that explains the data; the approach that seems to give the best explanation is a very good starting point for library design. Much of the literature on library design is directed towards absolute diversity, that is, designing a set of descriptors, an algorithm and a set of molecules that maximises the diversity of the molecules with respect to the known universe of drug-like molecules. This paper will not deal with this field, which has been well reviewed elsewhere [7, 8].

1.6 Reagent Filtering and Drug-likeness

The filtering of reagents for reactivity is not a major issue in the design of small and medium libraries, because there is a lesser degree of automation and manual intervention in the synthesis that can be performed if the compounds are thought to be key to the design. The issue is more often that there are too many choices, and the size of the virtual library precludes some of the design methods. However, it is vital not to lose sight of other considerations. As Higgs *et al.* [9] put it: "... [compounds] must not be so diverse as to be pharmaceutically unreasonable". They implemented a series of rules based on substructural queries that assign "demerits" to compounds. If any compound gains too many demerits, it is rejected. Others have also described a series of substructural filters designed to eliminate molecules containing toxic or very reactive substructures [10-12]. Care must be taken to apply these filters at the 'product' level, as reagents by definition contain reactive groups. This can be done by building a virtual library using a simple core with one substituent position and the list of reagents to be filtered, or by enumerating the library fully. Another suggestion is to order the reagents by their complexity (number of functional groups, rotatable bonds, degree of branching), so that the simplest are considered first. This is a sensible move, if the chemistry is difficult, or the design needs to produce ideas very quickly.

The next level of filters must be applied to the fully enumerated library, or the process is not meaningful. The "rule-of-5" criteria described by Lipinski *et al.* [13] is a quick and crude method for assessing bioavailability.

A compound is deemed to fail the rule-of-5 check if it possesses two or more of the following features:

- more than 5 hydrogen bond donors (i.e., N-H or O-H bonds)
- more than 10 hydrogen bond acceptors (i.e., any N or O, including those in donors)
- a ClogP value of greater than 5.0
- a molecular weight of greater than 500.0

The same concept has been used by Higgs *et al.* [9] in their demerit scheme, which seems to be a more sensible approach to adopt. If the SAR is only derived from weakly active compounds, then more leeway should be allowed in the characteristics of the library members: it may well be the case that a large fraction of the core structure, or one of the substitution positions is irrelevant and can be discarded, bringing the active compounds back within acceptable limits for the rule-of-5 test. If the active compounds have good activity, for example an IC50 of better than 100 nM, it makes more sense to include bioavailability criteria right from the start. Other factors that may influence this decision are the nature of the assay (receptor/cell-based/in vivo) and the maturity of the project. A very good review of the influence of oral bioavailability on the medicinal chemistry of combinatorial libraries has been performed by Mitscher *et al.* [14]. For more detailed research on the modelling of bioavailability, the reader is referred to the literature [15-19]. The experience of medicinal chemistry tells us that there is often a conflict between the design of compounds that have increased activity but that also have good ADME properties. If these factors can be included in the library design, then they should be, but with the understanding that these rules will tend to force the compounds (potentially the drug of the future) to conform to the conservatism of past medicinal chemists.

1.7 Enumeration of the virtual library

There are many methods available for enumerating libraries, mainly targeted at the bench chemist, but this does not seem to be appropriate, especially if the enumeration package does not interface easily with the design tools. Anecdotal evidence suggests that several modelling groups have written their own enumeration packages, because of dissatisfaction with the commercial offerings. There are two approaches to enumeration, fragment-based and reaction based. Fragment-based methods (for example ProjectLibrary[20]) take lists of reagents and clip them onto a core template to form the virtual library. This is very similar in concept to the Markush representation used in patents. However, libraries which are made by ring-forming reactions, or which contain variable stereochemistry do require a lot of care in their construction. There are many cases which cannot be done at all. The reaction-based builders (for example, SMIRKS from Daylight [21]) are absolutely general, but can be a little harder to set up. Fortunately, there is a limited number of reactions used in RPS today, so it is not too onerous for the modeller to set these reactions up for the chemists to use. Getting the stereochemistry right is not essential if only 2D molecular descriptors are being used to design the library, but it seems wasteful to discard this important information because of deficiencies in a piece of software.

The output from the enumeration should be a virtual library containing information about the molecular structure and a unique identifier as a minimum. Other useful information includes the reagents used to make a particular library member, and possibly their molecular weights (this can be used later in the purification of the synthesised library, to determine the level of impurities).

2. MOLECULAR DESCRIPTORS AND DIVERSITY METRICS

Most of the current literature on library design combines the molecular descriptors and the diversity metrics with the methods used to effect the design. In this chapter, the two topics are dealt with separately. Diversity measures and molecular descriptors are only tied artificially to a method by programmers. Ideally, one should be able to compute any descriptors or metrics one wants, and read that into the design program for analysis. In practise, import and export of data is never simple, but it should become so in the future. On the other hand, some descriptors and methods do work well together, and there is not always a good case for going outside the established paradigms.

2.1 Definitions

A molecular descriptor is a property of a single compound, for example, molecular weight. There are several thousand different descriptors that can be calculated from a knowledge of a molecular structure, of which some are more useful than others in the design of small and medium libraries. Rather than recapitulate descriptor properties and how they are calculated, the emphasis in this section will be on how these descriptors fit in with an SAR, continuing the theme of section 1.3. The first step is to try to produce an SAR hypothesis that explains as many of the experimentally measured activities as possible, and preferably in a form that is transferable to library design. The quality of the SAR can range from a simple binary classification into mainly actives and mainly inactives, to one in which absolute activity can be predicted with a reasonable degree of confidence. It should be stressed that the nature of the SAR will vary strongly on a case-by-case basis, so that there is no universal recipe. There are strong parallels in this approach to the concept of virtual screening [22-24], in that we are trying to find a method to estimate the activities of compounds before screening, or more generally, a method of estimating how much a compound will contribute to the healthy exploration of the SAR.

Molecules are compared using a similarity metric, which is a function of the molecular descriptors. This may involve direct comparison or implicit comparison via a ranking derived from an SAR model. Care has to be used when using similarity measures to compare dissimilar molecules. Most metrics have been validated by studying closely related compounds; there is no guarantee that they have any meaning at all for unrelated compounds. This issue has been elegantly discussed by Sello [25]. A diversity measure is a property of an ensemble of molecules, for example, the mean molecular weight of a virtual library, or the sum of the pairwise distances between all molecules in the ensemble.

2.2 2D Descriptors

2D descriptors are a property of the connectivity of a molecule, and hence contain little information about stereochemistry or local shape. Against that, they are easy to compute, store and manipulate. 2D descriptors have been a powerful weapon in setting up classical QSARs, perhaps because they can implicitly encode 3D features within a homologous series.

2.2.1 Substructural keys - SAR scenario: one active chemical family

Substructural keys encode information about the bonding patterns and functional groups present in a molecule. These keys have been very useful in clustering corporate databases in chemical families [26, 27]. The SAR scenario is that all the actives seem to belong to one chemical family, and their activities are weak. Techniques like determining the modal fingerprint can help to identify the features associated with activity [28]. The members of the virtual library can then be scored against the modal fingerprint, and the best members selected out. However, there is no guarantee that a high degree of match will translate to a more active compound, and therefore it is better to employ a minimum acceptable match threshold (one that gives a good separation of actives and inactives for the measured compounds), and use the library members that meet this criterion as a basis for further design. The argument here is that the SAR is weak, so that the emphasis in the design should be on exploring it further, and should therefore probably use features not in the present set of actives. For binary keys like the modal fingerprint, a number of similarity measures are available, including the Tanimoto coefficient (or more generally, the Tversky coefficient [29]) and the cosine coefficient [30]. This gives a measure of the distance between two objects. Two diversity metrics are relevant for library design: a relative score (distance to nearest active), which can be used to give a rank, or internal score (distance to the nearest molecule in the virtual library). The first metric can be used to focus the library design around the known actives, the second to ensure adequate exploration of the descriptor space. These metrics have been referred to as 'representative' and 'diversifying' metrics respectively. Research has been carried out as to which flavour of substructural fingerprint might be best to use [31].

2.2.2 Physicochemical properties - SAR scenario: several structurally unrelated compounds with weak activity

In this scenario, there are several structurally unrelated compounds, of which one or more is amenable to synthesis (and therefore expansion) by RPS. The commonality of these compounds is probably based around physicochemical properties, although this may be indicative of higher level similarities [11]. A useful starting point is therefore to begin with descriptors for hydrogen-bonding, polarity, lipophilicity and shape. If time and resources allow (remembering that the time taken to perform the design has to fit in with the schedules of the library synthesis), many other descriptors can also be computed and added into the analysis, for example topological indices [32], HOMO/LUMO energies, BCUT descriptors [33] or any QSAR descriptor available for the entire dataset. The claimed advantage of the BCUT descriptors over more traditional descriptors such as molecular weight, ClogP or topological indices, is that they reflect both molecular connectivities and atomic properties relevant to intermolecular interactions. Standard QSAR methods can then be employed to obtain a set of descriptors that classify actives and inactives. It is important to choose a method that allows one to score the virtual library, or to rank all the compounds within the library. A QSAR equation will give relative weightings for the different factors, eliminating those descriptors which are not relevant. An alternative method for setting up a diversity metric is to create a partitioned grid. The axes of the grid are the various descriptors, and with a suitable choice of descriptors, the actives should clump into one region of the partition grid. The χ^2 statistic can be used to test the degree of clumping.

$$\chi^2 = \Sigma(p_i - \langle p \rangle)^2/\langle p \rangle;\ \langle p \rangle - \Sigma p_i/N \qquad (1)$$

The diversity metric can then be the distance to the nearest object, or the degree of overlap between the two distributions. The second metric has to be calculated on the fly, as it is dependent on the membership of the distribution, which will vary according to which library subset is selected. Diversifying metrics based on molecular properties have been used to cluster corporate databases [34], and so are available to balance out the representative metric.

2.3 Single Conformation 3D descriptors

Several molecular descriptors can be computed from the 3D structure of a molecule. As a first approximation, it is assumed that a good low-energy structure will suffice to provide a reasonable benchmark for comparing molecules. This assumes that the molecules being compared will have similar binding modes, so that one is comparing like with like. The assumption is reasonable, as combinatorial chemistry is often centred around decorating a common core. If the core has a significant interaction with the receptor, or the lead series being explored is semi-mature and the changes being made are not radical, then it is likely that the binding mode will be the same.

2.3.1 Topomeric Descriptors - SAR scenario: exploration of an established lead.

Cramer and co-workers have proposed a bioisosteric descriptor based on the steric fields generated around a single reagent attached to a common core [35]. This would enable one to replace a fragment in a molecule with one which had a similar 3D shape or property, but a different structure. The key part of the procedure is the automated alignment of the different groups at a particular substitution position of the core. The steric field around each group can be computed using a standard Methyl probe, and the distance between two groups is the root sum of the squared differences of each pair of steric field values. In a comparison with 2D substructural keys, it was found that groups could have very different shapes while having similar substructures, and vice versa. The library design metrics are analogous to the distance-based metrics described above in section 2.2.2, based on comparison to the active compound(s) and between library members.

2.4 Multiple Conformation 3D descriptors

Multiconformational 3D descriptors are the most sophisticated descriptors of a molecule, as they encode not just one spatial arrangement of key molecular recognition elements, but also all the low-energy arrangements too. This should surmount the issue of the binding conformation not necessarily being the lowest energy conformation. Unfortunately, these descriptors require significantly more resources to compute, store and handle.

2.4.1 Property Matching - SAR scenario: a few structurally related compounds with reasonable activity.

Chapman has published an interesting method for comparing molecules based on multiple conformations [36]. The similarity between two chemical objects is measured using terms to describe shape, polarity and hydrogen-bonding, using Bohm's weighting scheme [37]. A chemical object is a molecule in a particular conformation, so that the measure can be used to look at the similarity between conformations of the same molecule. The terms are simple rms distance functions, for speed of calculation. Again, for speed of calculation, it is assumed that there is a common core which can be used as a reference to overlay the chemical objects; in this respect, there are analogies with the topomeric approach (section 2.3.1). The diversity function is:

$$D = \sum_{molecules} \sum_{conformations} \min(dissimilarity) \quad (2)$$

It was found that the method was biased towards flexible molecules, so that an entropy term was included, based on the number of rotatable bonds in a molecule. This approach could also be used to set up an SAR hypothesis based on the best consensus alignment, but once the alignment has been generated, it is likely that the SAR could be better handled using CoMFA. The appealing feature of this work is the introduction of shape, polarity and conformation space; its drawback is the heavy computational load.

A variation on this theme is provided by workers who have used steric similarity matches to explain SARs. In this work, relative activity is correlated by the shape or property similarity to the most active compound, using programs like ASP [38] or CoMFA [39, 40] to compute the pairwise similarities. The strength of this approach is that one can compare structurally dissimilar compounds. The weakness (as with any relative measure), is that one cannot extrapolate to guess whether a compound might be more active (the most similar compound to the reference compound is the reference compound itself).

2.4.2 Pharmacophores - SAR scenario: a few structurally unrelated compounds with reasonable activity.

A good working definition of a pharmacophore has been given as 'the ensemble of steric and electronic features which are necessary to ensure supramolecular interactions with a specific biological target structure' [41]. The ensemble also contains distance information, that is, the geometric pattern of the features. The features, or 'centres', are defined as hydrogen-bond donors, acceptors, aromatic ring centres, hydrophobes, bases and acids. The concept of a pharmacophore is turned into a molecular descriptor through the construction of a set of artificial pharmacophores, formed by dividing the distances between pharmacophoric groups into bins e.g. 2-3, 3-4, 4-5Å and so on. Each combination of features and distances is used to set a single bit in a binary key; it is this key that is the pharmacophoric descriptor. To get the key for a molecule, a full conformational analysis is performed, to see which pharmacophores the molecule can express. There are issues with this descriptor: a small rigid molecule will set fewer bits in the key than, say, a hexapeptide. This is why a penalty based on the number of accessible conformations is frequently used in conjunction with a pharmacophore key. They are also costly to compute and store. Methods for the generation and use of pharmacophore keys are discussed in more detail elsewhere [42]. In an analogous way to the modal fingerprint discussed in section 2.2.1, pharmacophore keys can be combined to give the common features of a set of active molecules as an overlap key, which encodes the SAR. Other methods for the determination of consensus pharmacophores are available [43]. The representative metric is then derived from the distance between the overlap key and the keys of the structures in the virtual library, either taken individually or as an overlap key for a sublibrary [44]. The diversifying metric is the number of different pharmacophores expressed by the sublibrary ensemble. An interesting variation on this theme is the development of frequency-based pharmacophore keys, which encode not just whether a pharmacophore is expressed, but also how often. This will allow the overlap key to encode the complete pharmacophore profile of the active compounds.

One issue with using pharmacophores as the scoring measure is that highly flexible molecules usually express more pharmacophores than rigid ones. Some workers claim this as an advantage and have proposed information-rich libraries of flexible molecules [45]. Personal experience suggests that flexible molecules are less desirable than their more rigid cousins due to issues of optimisation, lower hit rates, and bioavailability. The total number of pharmacophores present in the selected library, or the number of available conformations, can be used to weight the score against promiscuous molecules.

2.5 CoMFA - SAR scenario: an established SAR.

An established SAR can sometimes be described by a CoMFA model, which can be used to predict the activity of compounds, and to highlight the regions of space responsible for activity. Once the model has been obtained, a novel compound can be inserted into the model and its activity predicted, making this method appealing for scoring a library. However, as anyone who has constructed a CoMFA model will be aware, the correct alignment of the active molecules is vital to obtaining a good model, and the same is true for molecules whose activities are to be predicted. It is therefore important to construct a good rule for automated alignment, or the process for scoring a virtual library will be very tedious, and the scores may be unreliable. This was the issue tackled in the development of the topomeric descriptors (section 2.3.1). In contrast to the topomeric descriptors, the active compounds may not have a significant common core, or precisely the same mode of binding. One option to setting up the alignment would be to create a 3D database of the virtual library, and search this with a 3D query [46]. A CoMFA model will also highlight the regions of space that have not been adequately mapped by the SAR series (although the author is not aware of any easy method for quantifying how well a novel compound explores this space). The representative metric is therefore the CoMFA scores; the diversifying metric would be the root sum of the squared differences for each pair of CoMFA field values. Although the latter metric will take a large amount of time and memory to compute, it need only be done once.

2.6 Structure-Based Design - SAR scenario: a knowledge of the receptor site and binding mode

This scenario is not within the scope of this paper, other than to comment that the amount of information in the SAR, and its quality, is very high, so that careful design will often reap substantial rewards in terms of the number of design cycles required to obtain a highly active compound. This topic is discussed by Roe [47]. As for the CoMFA model, the representative metric is the score of the match of each compound to the receptor, the diversifying metric is a property difference. The key issues are the generation of an appropriate docking of the molecules to the receptor, and the speed and accuracy of the scoring of the interaction [48].

3. DESIGN STRATEGIES

This section discusses the different strategies used to select a subset of the library. It should be remembered that the best subset selection is neither arbitrarily nor maximally diverse [49], because we cannot guarantee a direct and strong link between the SAR and the diversity metrics used.

3.1 Random design

It may appear incongruous to have a section on random design on a paper dedicated to rational methods, but if the SAR is very shaky, and the size of the virtual library greatly exceeds the size of the proposed library, then a random selection of reagents that leads to drug-like molecules may be the most cost-effective way to proceed. Of course, the design will never be truly random, as a good medicinal chemist will favour certain reagents, perhaps subconsciously using a mental Topliss tree in the selection.

3.2 Design based on reagents

In *reagent-based* selection, the choice of a subset is made so as to maximize the scores of the reagents at each position without considering the reagents at the other positions, or the scaffold. A good example of such a method is that reported by the Chiron group [50]. Of course, almost any of the published techniques for diverse subset selection may also be applied to reagent-based selection of reagents. If one can safely assume that the groups at the substitution points about the core fragments are independent of the groups at another substitution point, then these design methods can be used.

3.3 Design Based on Products

Alternatively, a *product-based* scheme can be envisaged, in which reagents are selected at all positions simultaneously, so that the score of the generated products is maximized. This type of approach has been championed by Gillet *et al.* [51] and by Lewis *et al.* [52]. Finally, one may pick the most diverse set of products and then deconvolute to find the sets of reagents required to make that set. This kind of approach is sometimes called *cherry-picking*.

There are some advantages and disadvantages to each of these approaches and each may be appropriate in certain design situations. In general, the cherry-picking approach will result in the most diverse set of products; however, there is the disadvantage that this approach does not result in a synthetically efficient combinatorial library. Reagent-based selection is fast, since one is not considering the enumerated combinatorial products in the analysis and thus it may be suitable for instances where the enumerated virtual library is very large. However, experiments have shown that a product-based reagent selection approach gives superior diversity to a reagent-based method [51]. Van Drie and Lajiness report similar findings [53]. Balanced against this is the fact that most product-based schemes can only deal with enumerated libraries of the order of 100,000 molecules, a number that is easily attainable, particularly with more than two variable positions on the template. In practice, one is likely to need to combine the reagent-based and product-based approaches. The reagent based selection methods can be used to filter the initial reagent lists to a size where the virtual library becomes tractable for analysis by a product-based method.

4. DIVERSITY ALGORITHMS

The discussion in section 2 focused on how to combine descriptors and SAR analysis methods into a form that would allow a virtual library to be scored. These scores are the basis for the design, which looks to obtain a suitably optimal subset of the library. There are different ways of finding the (near-)optimal subset, which will now be discussed. A point that needs to be reiterated is that the scores are only approximate, and that the libraries do need to some extent to explore unknown SAR space, or face the danger of optimising into a dead-end. The types of information generated by the descriptors and SAR are: (i) distance matrices, in which each element is the distance between a pair of molecules; (ii) ranks and scores, in which each molecule is scored on its own merits; (iii) ensemble scores, in which the score is a property of the match of the sublibrary selected against some external SAR standard.

4.1 Maximising distance matrix scores

Some descriptors are best described by diversity metrics that focus on the distance between objects. The precise functional form of the metric varies, but a useful example is:

$$D = \max(\Sigma\Sigma d_{ij}) \quad (3)$$

The equation for D is trying to maximise the separation of the molecules in the library. Other variations would be to minimise D, in the case where d_{ij} was the distance to the nearest active molecule. Several algorithms are available for finding the optimal selection. The simplest approach uses a standard 'greedy' algorithm that adds the molecule that will most increase the value of D for the current set of molecules. The contribution of a molecule will depend on the molecules already selected, as all the pairwise distances between objects are being considered. Other methods, such as D-optimal design and experimental design [54] are available, but all tend to pick compounds from the edges of the component space, rather than taking a more even sampling [55]. Willett's group have looked extensively both at different definitions of D [56] and algorithms for optimising D [57]. In the former case, they concluded that it was impossible to identify any of the four definitions studied as being superior to the others. When comparing algorithms however, the MaxMin algorithm gave better results than the alternatives under study.

A general dissimilarity selection algorithm has been recently reported by Clark [58]. There is an adjustable parameter in the algorithm that controls the balance between representativeness and diversity. However, this does require a comparison metric that is meaningful when measuring distance between quite dissimilar objects. Other functions for maximising dissimilarity have been suggested by Hassan *et al.* [59].

4.2 Maximising rank scores

A rank score is a very useful means of relating a compound to the SAR. If combinatorial effiency is not an issue, then rank scores are very simple to deal with, as the function to be maximised is:

$$D = \max(\Sigma d) \quad (4)$$

The top scoring compounds are identified by a simple sort of the virtual library scores. The situation is more complex when trying to generate an efficient library, as by including the top-scoring compound in the library, one may also drag in a larger number of low scoring compounds made from the reagents that make up the best compound. There is also the issue of whether one wants to have a bias towards the best-scoring compounds, the highest average score or the highest median score. In algorithmic terms, the task of finding the best combinatorial library is equivalent to a linear programming task, and so may be tackled using methods from that area, including integer optimisation, genetic algorithms or simulated annealing. The first method will give the absolute best solution, whereas the other two are stochastic and so will give a solution close in value to the best in a much shorter time. The error in the scores between the global and near-global maxima found by stochastic means is generally much lower than the errors in the original scoring function, so it makes sense to use the stochastic methods. Application of genetic algorithms [60-62] and simulated annealing [63] to library design have been published

4.3 Ensemble scoring and distribution fitting

An ensemble score is created by comparing the overall property of a library against some external standard. As an example, one may wish to select a library with an even distribution of molecular weights between 250 and 500 Daltons. This can be achieved by using a partitioning algorithm and a χ^2 score (section 2.2.2). The external standard is read in as a frequency histogram of partitions and frequency values. The frequency histogram for the proposed library is calculated by assigning the molecules from the candidate library to the partitions of the input histogram. The restraint penalty is computed as the sum across the bins of the difference in the actual and expected values. If the input histogram is perfectly even, the score is a maximum when each bin is equally occupied (figure 3).

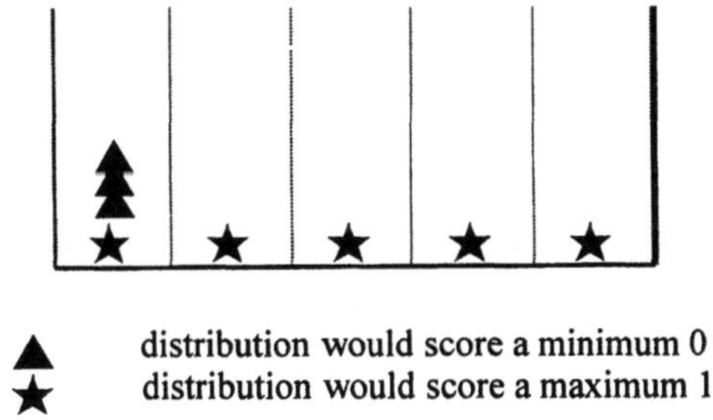

Figure 3. Partition scoring of properties. An even distribution (stars) gets the maximum score of one; the skewed distribution (triangles) gets the minimum score of zero.

As for ranked molecules, it is straightforward to get a good design for inefficient libraries, but for efficient libraries one again needs to use the techniques of simulated annealing or genetic algorithms described in the previous section.

4.4 Visualising diversity

One method for assessing whether a particular similarity metric has been able to classify actives and inactives apart, or whether a design library has similar characteristics to the set of known actives is to generate a visual representation. The issue is that often one is projecting a very high-dimensional space down to 2 or 3 dimensions. Sadowski *et al.* [64] have described the use of Kohonen neural networks to visualise the diversity of combinatorial libraries. Kohonen networks have the useful ability to project high-dimensional information down to just two dimensions, while ensuring that neighbouring points in the high dimensional space will still be neighbours in the lower dimensionality space. They projected a library based around cubane and one around xanthene onto the same Kohonen net, and were able to get an immediate visual picture of how similar the two libraries were to each other, and how diverse they were with respect to the property fingerprint (based on electrostatic potential).

5. COMBINATION OF DIVERSITY MEASURES

It is often the case that the virtual library can be scored by more than one method, for example, pharmacophore profile, physicochemical profile, SAR score, cost, drug-like properties and so on. A design decision must be taken as to how these scores should be weighted against each other. In the same way, the balance between optimising the representative metric and the diversifying metric needs to be set according to the maturity of the SAR and the needs of the lead optimisation strategy. It might be argued that the relative weighting of the factors should have been fixed as part of the determination of the SAR, but this is not always possible. A particular case in point might be trying to combine the overlap to pharmacophore key with a ClogP profile. There is no universal prescription for how to perform this weighting, other than to perform several designs with different weightings and analyse them in terms of any grey information that might be available about the SAR or the ease of synthesis. Lewis and co-workers have looked at this issue using the HARPick program [63].

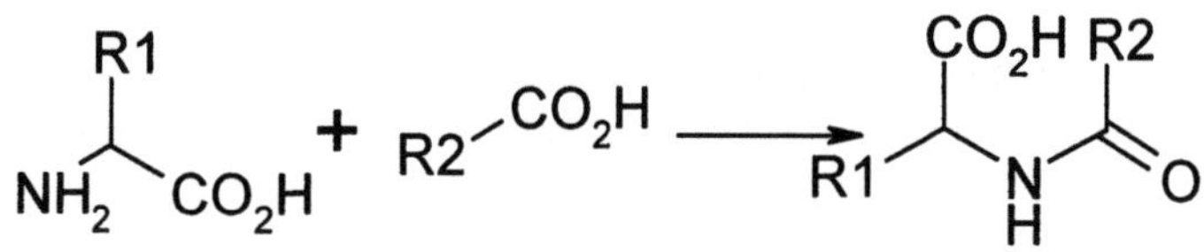

Figure 4. Two component hypothetical library used in the HARPick experiments, comprising "amino acid" (component 1) and acid (component 2) reagents

Two runs were undertaken, with the calculation constrained to select 20 amino acids and 50 acids from the simple amide bond forming reaction shown in figure 4. Three experiments were performed (i) maximize the internal pharmacophore diversity only (the diversifying metric) (ii) maximize internal pharmacophore diversity while trying to fill voids (the representative metric) in the SDF library [65] (iii) select the library randomly. The first run gave a library design that contained 84811 independent pharmacophores; constraining the design to minimise overlap with the SDF dropped the number of pharmacophores slightly to 76920. Random selections gave an average of 47344 pharmacophores in the design. These experiments show that design can produce much more diversity than random selection, but also that there is a tension between following an SAR and exploring diversity.

In another experiment, the goal was to select a set of 372 compounds from the SDF using different design criteria (i) maximize the number of pharmacophores only; (ii) maximize the number of pharmacophores while maximizing the shape partition scores (which have values between 0 and 1, 1 being the optimum); (iii) maximize the number of pharmacophores while maximizing the shape partition scores and minimizing the flexibility expressed as the number of conformers. The results are shown in Table 1. It can be seen that inclusion of partition scores gives a better spread of an external variable (for example, number of heavy atoms), even though the distribution of the variable in the SDF is strongly skewed (Figure 5). By contrast, the pharmacophore diversity is reduced. When flexibility is added in, the number of conformers decreases 200-fold, while the number of unique pharmacophores is not strongly affected. The use of these extra contraints will result in a more drug-like library being designed.

Table 12. Results from the selection of 372 diverse compounds from the SDF database

Calculation	Pharmacophores	Conformers	Shape Score 1	Shape 2	Shape 3
Run (i)	105222	4.0×10^8	0.97	0.69	0.70
Run (ii)	61913	6.4×10^8	1.00	0.90	0.85
Run (iii)	70656	2.4×10^6	0.90	0.79	0.54

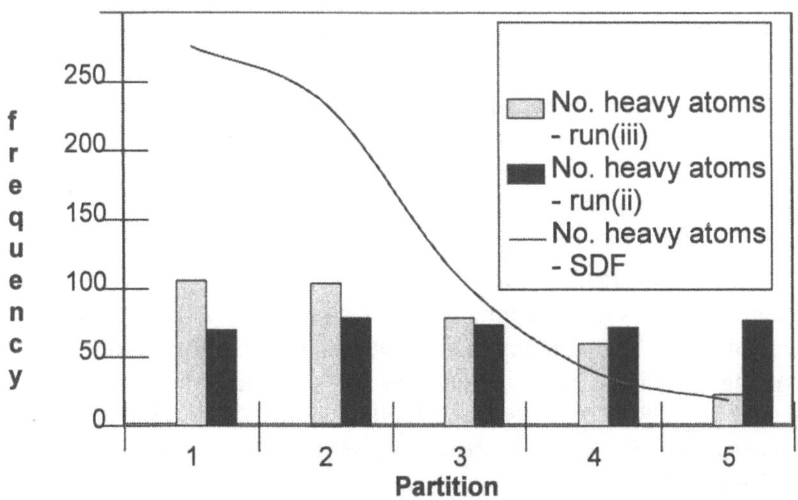

Figure 5. The graph of occupation frequency against the partitions of the number of heavy atoms in the molecules. The spread of occupancy is far more even for the HARPick runs than in the SDF database itself

Another experiment looked at the effect of changing the weighting of a factor. A virtual library was built from 100 amines and 100 acids, and HARPick was used to select a sublibrary of 400 amides in a combinatorially efficient manner. The design criterion was to maximise the unique/total pharmacophores while minimizing the cost of reagents. Table 2 shows the effect of increasing the weight of the reagent cost. In this case, it appears that the diversifying metric is not strongly affected by the constraint metric, but there is some negative interplay between them.

Table 13. Results from an experiment to investigate the effect of cost on the number of pharmacophores in a designed library

Cost Weight	Unique Pharmacophores	Total Pharmacophores	Total Cost
0.0	21230	171709	2238
0.1	21184	177073	1253
0.25	21003	184237	1049
0.33	21043	192570	954

6. COMMERCIAL PROGRAMS FOR LIBRARY DESIGN

This section will review briefly the current commercial offerings [66], in the sure knowledge that the reviews will be overtaken by the activities of the companies in this field. The objective is to look at the general strategy, rather than focus on the extra pieces of functionality which could be added to or improved at a later date.

6.1.1 Molecular Simulations

The library design modules in Cerius2 (v4.0 [67]) provide several useful functions for library design. The design strategy is built around molecular spreadsheets, so that many 2D and some single conformation-3D descriptors can be calculated for a virtual library. Cerius2 does have facilities for generating QSAR models from 2D descriptors, Molecular Field Analysis or pharmacophore hypothesis generation, so it is possible to generate SAR rankings which can be fed directly into the library design. In addition, the design strategy is cherry-picking by maximising a diversity score. There is also the opportunity to utilise the functions in Insight for structure-based drug design (for example, Ludi) to generate binding scores. MSI have also set up a consortium to investigate further descriptors, including pKa and logD, both of which are very important in describing bioavailability.

6.1.2 Tripos

The initial drive of Tripos's [68] first products in this field seemed to be towards the analysis, design and management of large, generally diverse libraries. The company have also been active in deriving new descriptors (section 2.3.1), and in trying to determine and validate a best set of descriptors to describe absolute diversity [69]. Modules are available for deriving SARs (particularly the CoMFA method), so the basic tools for focused library design are available, if not fully integrated yet. Tripos also supply the DVS software.

6.1.2.1 DVS

The DVS suite of programs written by Pearlman and co-workers [33] use the BCUT descriptors and the partitioning approach to design of small libraries. First the appropriate descriptors are defined by an analysis of the SAR, then the partitioning algorithms can be used to identify members of a virtual library that fill voids or occupy the same regions of diversity space as active compounds. This has the advantage over the experimental design methods of covering space in an even or in a directed fashion. This approach is only workable in very low dimensional diversity spaces. A space described by five descriptors, with 5 divisions per descriptor would lead to 3125 partitions, which is starting to move outside our synthetic scope. Against this, there is a situation in which large numbers of partitions are not a disadvantage, that is, the interpretation of results from an HTS run. In this case, the partitioning would be chosen to try to group all the actives together in a small region of space. The RPS library would then be designed to try to cover this region of space exclusively. In earlier versions of DVS, the combinatorial efficiency issue was not really handled well, other than by analysing which reagents occurred most frequently in the selected set. It is anticipated that later versions will address this issue. It should also be added that DVS also includes a tool (CombiDBMaker) for building libraries, which is quite easy to use, and which specifically addresses the stereochemistry issue, building on the foundations of the Concord and stereoplex programs.

6.1.3 Chemical Design

The ChemDiverse software for library profiling [70, 71] provides a number of useful tools for diversity assessment and, as it is an integral part of ChemX, can be linked directly to SAR models of all types. The ChemDiverse protocol for molecular diversity is based on trying to obtain the maximum coverage of pharmacophore space from the potential combinatorial chemistry products. The strategy chosen is cherry-picking. It is not possible as yet to modify the search criterion to include restraints such as shape, but it is possible to constrain designs through the use of upper and lower bounds for given properties.

7. PUBLISHED APPLICATIONS

The number of papers on the design of small RPS libraries driven by previous SAR is starting to increase as the tools for library design become more robust. Tropsha *et al.* [72] have looked at Bradykinin potentiating peptides. In this work, an SAR was derived from a training set of 28 peptides, and a genetic algorithm was used to select a sublibrary of pentapetides based on the predicted activity. A degree-of-fit term was used to prevent the selection of peptides that were greatly structurally different from the training set, to reduce errors arising from excessive extrapolation of the SAR. This approach is a cherry-pick method, as the end product is a linear polypeptide. Lui *et al.* [73] used a CoMFA-type approach to design analogues of (-)huperzine, an acetylcholine esterase inhibitor. The analogues were scored by similarity to (-)huperzine, and the quality of the selected library was evaluated by comparison to the x-ray structure of huperzine and acetylcholine esterase. At least one known active analogue was found by this method. There are several other papers which include in their keywords the terms 'combinatorial chemistry' and 'library design', but these seem to be concerned more with combinatorial explosion around a central motif [74] than with development and exploitation of a SAR.

8. CONCLUSIONS

The use of focused combinatorial or RPS libraries is becoming an important weapon in lead exploration and optimisation. The challenge for the modelling community is to harness the experience and knowledge we have in generating SARs for use in library design. This paper has described several strategies by which this objective can be achieved. The results now appearing in the literature show the value of designing focused libraries, and much more progress can be expected in this field in the years to come.

ACKNOWLEDGMENTS

I would like to thank my colleagues at Eli Lilly and at Rhone-Poulenc Rorer, in particular David Clark, Stephen Pickett and Ian Watson, for many stimulating discussions.

REFERENCES

1. Brown, R.D. Descriptors for diversity analysis. *Perspect. Drug Disc. Des.*, 1997, **7/8**, 31-39.
2. Dooley, C.T., Ny, P., Bidlack, J.M. and Houghten, R.A. Selective ligands for the sigma, delta, and kappa opioid receptors identified from a single mixture based tetrapeptide positional scanning combinatorial library. *J. Biol. Chem.*, 1998, **273**, 18848-18856.
3. Combs, A.P., Kapoor, T.M., Feng, S.B., Chen, J.K., Daudesnow, L.F. and Schreiber, S.L. Protein Structure-Based Combinatorial Chemistry: Discovery of Non-Peptide Binding Elements To Src SH3 Domain. *J. Am. Chem. Soc.*, 1996, **118**, 287-288.
4. Felder, E.R. and Poppinger, D. Combinatorial Compound Libraries for Enhanced Drug Discovery Approaches. *Advances in Drug Research,* 1997, **30**, 111-199.
5. Taylor, R. Simulation Analysis of Experimental Design Strategies for Screening Random Compounds as Potential New Drugs and Agrochemicals. *J. Chem. Inf. Comput. Sci.*, 1995, **35**, 59-67.
6. Kubinyi, H. Similarity and dissimilarity: a medicinal chemist's view. *Perspect. Drug Disc. Des.* 1998, **9/10/11**, 225-252.
7. Thompson, L.A. and Ellman, J.A., Synthesis and Applications of Small Molecule Libraries. *Chem. Rev.*, 1996, **96**, 555.
8. Gordon, E.M., Gallop, M.A. and Patel, D.V., Strategy and Tactics in Combinatorial Organic Synthesis. Applications to Drug Discovery. *Acc. Chem. Res.*, 1996, **29**, 144.
9. Higgs, R.E., Bemis, K.G., Watson, I.A. and Wikel, J.H. Experimental Designs for Selecting Molecules from Large Chemical Databases. *J. Chem. Inf. Comput. Sci.* 1997, **37**, 861-870.
10. Lajiness, M. Evaluation of the Performance of Dissimilarity Selection Methodology. In *QSAR: Rational Approaches to the Design of Bioactive Compounds,* Eds Silipo, C. and Vittoria, A., Escom, **1991**, pp. 201-204.
11. Lewis, R.A., Mason, J.S. and McLay, I.M. Similarity Measures for Rational Set Selection and Analysis of Combinatorial Libraries: The Diverse Property-Derived (DPD) Approach. *J. Chem. Inf. Comput. Sci.* 1997, **37**, 599-614.
12. Rishton, G.M. Reactive Compounds and In Vitro False Positives in HTS. *Drug Discovery Today,* 1997, **2**, 382-384.
13. Lipinski, C.A., Lombardo, F., Dominy, B.W. and Feeney, P.J. Experimental and Computational Approaches to Estimate Solubility and Permeability in Drug Discovery and Development Settings. *Adv. Drug Delivery Rev.* 1997, **23**, 3-25.
14. Fecik, R.A., Frank, K.E., Gentry, E.J., Menon, S.R., Mitscher, L.A. and Telikepalli, H. The search for orally active medications through combinatorial chemistry. *Med. Res. Rev.*, 1998, **18**, 149-185.
15. van de Waterbeemd, H., Camenisch, G., Folkers, G. and Raevsky, O.A. Estimation of Caco-2 Cell Permeability using Calculated Molecular Descriptors. *QSAR,* 1996, **15**, 480-490.
16. Palm, K., Luthmann, K., Ungell, A.-L., Strandlund, G. and Artursson, P. Correlation of Drug Absorption with Molecular Surface Properties. *J. Pharm. Sci.*, 1996, **85**, 32-39.
17. Palm, K., Artursson, P. and Luthmann, K. Experimental and Theoretical Predictions of Intestinal Drug Absorption. In *Computer-Assisted Lead Finding and Optimization: Current Tools for Medicinal Chemistry,* Eds van de Waterbeemd, H., Testa, B. and Folkers, G., Wiley-VCH, Weinheim, 1997, pp. 277-289.

18. Palm, K., Stenberg, P., Luthmann, K. and Artursson, P. Polar Molecular Surface Properties Predict the Intestinal Absorption of Drugs in Humans. *Pharm. Res.*, 1997, **14**, 568-571.
19. Rodrigues, A.D. Preclinical Drug Metabolism in the Age of High-Throughput Screening: An Industrial Perspective. *Pharm. Res.,* 1997, **14**, 1504-1510.
20. Molecular Design Ltd, San Leandro, CA94577, USA.
21. Daylight Chemical Information Systems, Mission Viejo, CA92691, USA.
22. Zheng, Q. and Kyle, D.J. Computational Screening of Combinatorial Libraries. *BioOrg. Med. Chem.,* 1996, **4**, 631-638.
23. Walters, W.P., Stahl, M.T. & Murcko, M.A. Virtual Screening - an overview. *Drug Discovery Today*, 1998, **3**, 160-178.
24. Zheng, Q. and Kyle, D.J. Computational Screening of Combinatorial Libraries via Multicopy Sampling. *Drug Discovery Today,* 1997, **2**, 229-234.
25. Sello, G. Similarity measures: Is it possible to compare dissimilar structures? *J. Chem. Inf. Comput. Sci.*, 1998, **38**, 691-701.
26. Downs, G.M. and Willett, P. Similarity Searching in Databases of Chemical Structures. *Rev. Comp. Chem.,* 1996, **7**, 1-66.
27. Barnard, J.M. and Downs, G.M. Chemical Fragment Generation and Clustering Software. *J. Chem. Inf. Comput. Sci.,* 1997, **37**, 141-142.
28. Shemetulskis, N.E., Weininger, D., Blankley, C.J., Yang, J.J. and Humblet, C. Stigmata: An Algorithm to Determine Structural Commonalities in Diverse Datasets. *J. Chem. Inf. Comput. Sci.,* 1996, **36**, 862-871.
29. Bradshaw, J. Introduction to Tversky similarity measure. Available at URL: http://www.daylight.com/meetings/mug97/Bradshaw/MUG97/tv_tversky.html
30. Turner, D.B., Tyrrell, S.M. and Willett, P. Rapid Quantification of Molecular Diversity for Selective Database Acquisition. *J. Chem. Inf. Comput. Sci.,* 1997, **37**, 18-22.
31. Downs, G.M. and Barnard, J.M. Techniques for Generating Descriptive Fingerprints in Combinatorial Libraries. *J. Chem. Inf. Comput. Sci.,* 1997, **37**, 59-61.
32. Kier, L.H. and Hall, L.B. The Molecular Connectivity Chi Indexes and Kappa Shape Indexes in Structure-Property Modeling. *Rev. Comp.Chem.,* 1991, **2**, 367-422.
33. Pearlman, R.S.. Stewart, E.L., Smith, K.M. and Balducci, R. Novel Software Tools for Combinatorial Chemistry and Chemical Diversity. Paper given at the 1997 Charleston Conference *Advancing New Lead Discovery*, Isle of Palms, SC U.S.A. (March 1997).
34. Downs, G.M., Willett, P. and Fisanick, W. Similarity Searching and Clustering of Chemical Structure Databases Using Molecular Property Data. *J. Chem. Inf. Comput. Sci.,* 1994, **34**, 1094-1102.
35. Cramer, R.D., Clark, R.D., Patterson, D.E. and Ferguson, A.M. Bioisosterism as a Molecular Diversity Descriptor: Steric Fields of Single Topomeric Conformers. *J. Med. Chem.,* 1996, **39**, 3060-3069.
36. Chapman, D. The Measurement of Molecular Diversity: A Three-Dimensional Approach. *J. Comput.-Aided Mol. Des.*, 1996, **10**, 501-512.
37. Bohm, H.J. The development of a simple empirical scoring function to estimate the binding constant for a protein-ligand complex of known three-dimensional structure. *J. Comput.- Aided Mol. Des.,* 1994, **8**, 243-256.
38. Good, A.C. & Richards, W.G. Rapid evaluation of shape similarity using Gaussian functions. *J. Chem. Inf. Comput. Sci.*, 1993, **33**, 112-116.
39. Cramer, R.D., DePriest, S.A., Patterson, D.E. and Hecht, P. The Developing Practice of Comparative Molecular Field Analysis. In *3D QSAR in Drug Design*. Ed. Kubinyi, H., ESCOM, Leiden, 1993, pp. 443-485.

40. Kim, K.H. Comparative molecular field Analysis (CoMFA). In *Molecular Similarity in Drug Design*. Ed.: Dean, P.M., Chapman & Hall: London 1995, pp 291-331.
41. Wermuth, C. and Langer, T. (1993) Pharmacophore Identification. In: *3D QSAR in Drug Design*, Ed. Kubinyi, H., ESCOM, Leiden, 1993, pp. 117-136.
42. Mason J. S. Absolute versus Relative Similarity and Diversity. In: *Molecular Diversity in Drug Design*, Ed. Dean P.M. and Lewis R.A., Kluwer, 1999, Ch. 4.
43. VanDrie, J.H. Strategies for the determination of pharmacophoric 3D database queries, *J. Comput.-Aided Mol. Des.*, 1997, **11**, 39-52.
44. Pickett, S.D., Luttmann, C., Guerin, V., Laoui, A. and James, E. DIVSEL and COMPLIB - Strategies for the Design and Comparison of Combinatorial Libraries using Pharmacophoric Descriptors. *J. Chem. Inf. Comput. Sci.*, 1998, **38**, 144-150.
45. Myers, P.L., Green, J.W., Saunders, J. and Teig, S.L. Rapid, Reliable Drug Discovery. *Today's Chemist at Work* 1997, July/August, 46-53.
46. Good, A.C. and Mason, J.S. Three-Dimensional Structure Database Searches. *Rev. Comp. Chem.*, 1996, **7**, 67-118
47. Roe D.C. Molecular diversity in site-focused libraries. In: *Molecular Diversity in Drug Design*, Ed. Dean P.M. and Lewis R.A., Kluwer, 1999, Ch. 7.
48. Eldrige, M. D., Murray, C.W., Auton, T.R., Paolini, G.V. and Mee, R.P. Empirical scoring functions: I. The development of a fast empirical scoring function to estimate the binding affinity of ligands in receptor complexes. *J. Comput.-Aided Mol. Des.*, 1997, **11**, 425-445.
49. Blaney, J.M. and Martin, E.J. Computational Approaches for Combinatorial Library Design and Molecular Diversity Analysis. *Curr. Opin. Chem. Biol.*, 1997, **1**, 54-59.
50. Martin, E.J., Blaney, J.M., Siani, M.A., Spellmeyer, D.C., Wong, A.K. and Moo, W.H. Measuring Diversity: Experimental Design of Combinatorial Libraries for Drug Discovery. *J. Med. Chem.*, 1995, **38**, 1431-1436.
51. Gillet, V.J., Willett, P. and Bradshaw, J. The Effectiveness of Reactant Pools for Generating Structurally-Diverse Combinatorial Libraries. *J. Chem. Inf. Comput. Sci.*, 1997, **37**, 731-740.
52. Lewis, R.A., Good, A.C. and Pickett, S.D. Quantification of Molecular Similarity and Its Application to Combinatorial Chemistry. In *Computer-Assisted Lead Finding and Optimization: Current Tools for Medicinal Chemistry* Eds. van de Waterbeemd, H., Testa, B. and Folkers, G., Wiley-VCH:Weinheim, 1997, pp. 135-156.
53. van Drie, J.H. and Lajiness, M.S. Approaches to Virtual Library Design. *Drug Discovery Today*, 1998, **3**, 274-283.
54. Ross, T.M. and Reitz, A.B. Measuring Diversity: Experimental Design of Combinatorial Libraries for Drug Discovery. *Chemtracts: Organic Chemistry*, 1996, **9**, 110-114.
55. Agrafiotis, D.K. Stochastic Algorithms for Maximizing Molecular Diversity. *J. Chem. Inf. Comput. Sci.*, 1997, **37**, 841-851.
56. Holliday, J.D. and Willett, P. Definitions of Dissimilarity for Dissimilarity-Based Compound Selection. *J. Biomolecular Screening*, 1996, **1**, 145-151.
57. Snarey, M., Terrett, N.K., Willett, P. and Wilton, D.J. Comparison of algorithms for dissimilarity-based compound selection. *J. Mol. Graph. Modelling*. In the press.
58. Clark, R.D. OptiSim: An Extended Dissimilarity Selection Method for Finding Diverse Representative Subsets. *J. Chem. Inf. Comput. Sci.*, 1997, **37**, 1181-1188.
59. Hassan, M., Bielawski, J.P., Hempel, J.C. and Waldman, M. Optimisation and Visualisation of Molecular Diversity of Combinatorial Libraries. *Molecular Diversity*, 1996, **2**, 64-74.

60. Weber, L., Wallbaum, S., Broger, C. and Gubernator, K. Optimisation of the Biological Activity of Combinatorial Compound Libraries by a Genetic Algorithm. *Angew. Chem. Int. Ed. Eng.*, 1995, **34**, 2280-2282.
61. Sheridan, R.P. and Kearsley, S.K. Using a Genetic Algorithm to Suggest Combinatorial Libraries. *J. Chem. Inf. Comput. Sci.*, 1995, **35**, 310-320.
62. Weber, L. And Almstetter, M. Diversity in Very Large Libraries. In: *Molecular Diversity in Drug Design*, Ed. Dean P.M. and Lewis R.A., Kluwer, 1999, Ch. 5.
63. Good, A.C. and Lewis, R.A. New Methodology for Profiling Combinatorial Libraries and Screening Sets: Cleaning Up the Design Process with HARPick. *J. Med. Chem.*, 1997, **40**, 3926-3936.
64. Sadowski, J., Wagener, M. And Gasteiger, J., Assessing similarity and diversity of combinatorial libraries by spatial autocorrelation functions and neural networks, *Angew. Chem. Int. Ed. Eng.*, 1995, **34**, 2674-2677.
65. Standard Drug File, (now known as the World Drug Index), Derwent Publications Ltd. 14 Great Queen Street, London, WC2B, UK.
66. Warr, W.A. Commercial Software Systems for Diversity Analysis. *Perspect. Drug Disc. Des.*, 1997, **7/8**, 115-130.
67. Molecular Simulations Inc., San Diego CA 92121, USA.
68. Tripos, St. Louis, Missouri 63144, USA.
69. Patterson, D.E., Cramer, R.D., Ferguson, A.M., Clark, R.D. and Weinberger, L.E. Neighbourhood Behaviour: A Useful Concept for Validation of Molecular Diversity Descriptors. *J. Med. Chem.*, 1996, **39**, 3049-3059.
70. Davies, E.K. Using Pharmacophore Diversity to Select Molecules to Test from Commercial Catalogues. In *Molecular Diversity and Combinatorial Chemistry: Libraries and Drug Discovery*. Eds: Chaiken, I.M. and Janda, K.D. ACS:Washington DC, 1996, pp 309-316.
71. Chemical Design Ltd, part of OMG, Oxford Science Park, Oxford OX4 4GA, UK.
72. Cho, S.J., Zheng, W. and Tropsha, A. Rational Combinatorial Library Design. 2. Rational Design of Targeted Combinatorial Peptide libraries using chemical similarity probe and inverse QSAR approaches. *J. Chem. Inf. Comput. Sci.*, 1998, **38**, 259-268.
73. Lui, D., Jiang, H., Chen, K. and Ji, R. A new approach to design virtual combinatorial library with genetic algorithm based on 3D property. *J. Chem. Inf. Comput. Sci.*, 1998, **38**, 233-242.
74. Brown, P.J., Smith-Oliver, T.A., Charifson, P.S., Tomkinson, N.C.O., Fivush, A.M., Sternbach, D.D., Wade, L.E., Orband-Miller, L., Parks, D.J., Blanchard, S.G., Kliewer, S.A., Lehmann, J.H. and Willson, T.M. Identification of Peroxisome Proliferator-activated Receptor Ligands from a Biased Chemical Library. *Chemistry and Biology*, 1997, **4**, 909-918.

Index

MIX
Papier aus verantwortungsvollen Quellen
Paper from responsible sources
FSC® C105338

FSC
www.fsc.org

If you have any concerns about our products,
you can contact us on
ProductSafety@springernature.com

In case Publisher is established outside the EU,
the EU authorized representative is:
Springer Nature Customer Service Center GmbH
Europaplatz 3, 69115 Heidelberg, Germany

Printed by Libri Plureos GmbH
in Hamburg, Germany